WOMEN IN ENGINEERING AND SCIENCE
Employers policies and practices

Informing policy by establishing facts

The Policy Studies Institute (PSI) is Britain's leading independent research organisation undertaking studies of economic, industrial and social policy, and the workings of political institutions.

PSI is a registered charity, run on a non-profit basis, and is not associated with any political party, pressure group or commercial interest.

PSI attaches great importance to covering a wide range of subject areas with its multi-disciplinary approach. The Institute's 40+ researchers are organised in teams which currently cover the following programmes:

Family Finances and Social Security
Health Studies and Social Care
Innovation and New Technology
Quality of Life and the Environment
Social Justice and Social Order
Employment Studies
Arts and the Cultural Industries
Information Policy
Education

This publication arises from the Employment Studies programme and is one of over 30 publications made available by the Institute each year.

Information about the work of PSI, and a catalogue of available books can be obtained from:

Marketing Department, PSI
100 Park Village East, London NW1 3SR

WOMEN INTO ENGINEERING AND SCIENCE

Employers' policies and practices

**Susan McRae, Fiona Devine
and Jane Lakey**

POLICY STUDIES INSTITUTE
100 PARK VILLAGE EAST, LONDON NW1 3SR
Tel: 071-387 2171 Fax: 071-388 0914

**The publishing imprint of the independent
POLICY STUDIES INSTITUTE
100 Park Village East, London NW1 3SR
Telephone: 071-387 2171; Fax: 071-388 0914**

© Policy Studies Institute 1991

All rights reserved. No part of this publication may be reproduced, stored in a retrieval system or transmitted, in any form or by any means, electronic or otherwise, without the prior permission of the Policy Studies Institute.

ISBN 0 85374 519 6

A CIP catalogue record of this book is available from the British Library.

1 2 3 4 5 6 7 8 9

How to obtain PSI publications

All book shop and individual orders should be sent to PSI's distributors:

BEBC Ltd
9 Albion Close, Parkstone, Poole, Dorset, BH12 2LL

Books will normally be despatched in 24 hours. Cheques should be made payable to BEBC Ltd.

Credit card and telephone/fax orders may be placed on the following freephone numbers:

FREEPHONE: 0800 262260 FREEFAX: 0800 262266

Booktrade Representation (UK & Eire)

Book Representation Ltd
P O Box 17, Canvey Island, Essex SS8 8HZ

PSI Subscriptions

PSI Publications are available on subscription.
Further information from PSI's subscription agent:

Carfax Publishing Company Ltd
Abingdon Science Park, P O Box 25, Abingdon OX10 3UE

Laserset by Policy Studies Institute
Printed in Great Britain by BPCC Wheatons Ltd, Exeter

Acknowledgements

The study on which this report is based was funded by the ESRC/SERC Joint Committee and the Employment Department (formerly Training Agency). Any opinions expressed are those of the authors, and do not necessarily represent the views of the Joint Committee or the Employment Department. We are very grateful for their support.

The study would not have been possible without the cooperation of the ten companies that agreed to participate in the research. PSI is very grateful to the many people who spent much time and energy helping us with our enquiries. The companies that participated in the study were as follows:

Boots The Chemist plc
British Gas plc
British Aerospace plc
Esso UK plc
ICI Group
Marconi Defence Systems Limited (Stanmore)
Ove Arup Partnership
The Plessey Company Limited
STC Telecommunications
Westland Helicopters Limited

Contents

1 Introduction	1
Technological change and occupational trends	2
Methods of research	5
The structure of the report	6
Part 1 Recruiting women into engineering and science	8
2 Schools liaison	12
Putting policy into practice	13
Encouraging young women into engineering and science	17
The role of teachers	19
Pupils' responses to schools liaison	20
Company evaluation of schools liaison initiatives	23
The future of schools liaison	24
3 Graduate recruitment	26
Recruiting graduates	26
Attracting women graduates: the employee view	32
Equal opportunities policy	33
Interviewing potential recruits	40
Monitoring	44

Part 2 Retention policy and practice 47

4 Training 49
The provision of training 49
The allocation of training 53
Satisfaction with training 55
Equal opportunities for training 56

5 Career progression 58
Opportunities for promotion 58
Assessment and career progression 62
Criteria for promotion: the employees' views 65
Equal opportunities for promotion 70

6 Retaining women engineers and scientists 79
Returning to work after childbirth 80
Help with childcare 82
Maternity leave 84
Career-break schemes 85
Part-time employment 89
The impact of career-break schemes and part-time working on women's promotion prospects 92

7 Conclusions and recommendations 96
Schools liaison 96
Graduate recruitment 99
Training 102
Career progression 103
The retention of women engineers and scientists 106

Appendix
Characteristics of the employees who participated in the study 109

Bibliography 112

1 Introduction

British industry has experienced shortages of qualified engineers and scientists for many decades. Commissioned by the then Secretary of State for Industry to conduct an inquiry into the engineering profession in 1977, Finniston reported that *there are undoubtedly serious shortages of engineers in some areas of industry* (HMSO 1980). Finniston went on to note that in some measure these shortages were due to inadequate deployment of the current stock of engineers, but against this, suggested that shortages were probably underestimated by employers as a result of their failure to appreciate the potential benefits to be gained from the more widespread use of qualified engineers. Finniston recommended two courses of action: first, to produce as many engineers as possible; and secondly, to ensure that employers make optimum use of the engineers available. It has become increasingly recognised that the fulfilment of these two goals requires that better use is made of women as engineers, scientists and technologists.

In this report, we examine the policies and practices of ten major British companies in seeking to attract, recruit and retain women in scientific and engineering occupations. During the 1980s, employers increasingly had difficulties recruiting sufficient numbers of either men or women for these occupations. Accordingly, although our first concern is with company policies and practices governing the employment of women, we also consider corporate recruitment and retention policies and practices more broadly. Thus, our findings have wider applicability than studies which focus exclusively upon one group in the labour market and will be relevant to organisations, individuals and policy-makers generally concerned with problems of labour supply.

From the perspective of the firm, ensuring an adequate labour supply is a single problem with a range of possible solutions. The solutions undertaken by employers as regards the supply of qualified women engineers and scientists have been various and include participation in schools liaison projects aimed at encouraging schoolgirls to choose subjects which later will allow them to take up technical careers. Companies also have sponsored women graduates in the expectation that such women will seek employment with them after graduation. They have sought to retain qualified women employees in their organisations through the introduction of career breaks, extended maternity leave and part-time working arrangements, and have brought in equal opportunities policies to ensure that optimal use is made within the organisation of women employees through the removal of barriers in recruitment, training and promotion.

Our report evaluates the effectiveness of these solutions from the perspective of managers and employees alike. The evaluation is undertaken from the perspective of the firm on the expectation that employer-led initiatives may be a more effective source of change in comparison with a tradition of waiting for educational or cultural change. A large part of the value of such an evaluation lies in the opportunities it presents for the widespread dissemination of employers' good practices in recruiting and retaining valued personnel.

Technological change and occupational trends

The increasing pace of technological change since the early 1980s has been documented by numerous commentators (Daniel 1987; Gershuny 1983; Lane 1988; Martin 1988). Particular attention has been paid to the introduction of microelectronic equipment on both the shop floor and in offices as the computer has come to aid a wide variety of work tasks from the production process to marketing, sales and administration. Technological developments of this kind are not just a matter of introducing new machinery into existing workplaces. Decisions also must be made about the way in which new machinery is to be introduced, the extent to which its facilities will be used and by whom, and what new skills might be required (McLoughlin and Clark 1988). Of particular relevance to the present study are changes in employment opportunities resulting from technological developments. The introduction and use of new technology frequently has been assumed to entail widespread job loss (Jenkins and Sherman 1979). But as Freeman and Soete (1987) have noted, the introduction of new technology can mean enhanced employment opportunities for those with technical skills. In practice, moreover, the greatest impact of technological development has been on the category of people recruited (Daniel 1987; Northcott 1986).

The spread of new technologies has tended to be concentrated in a small number of areas, particularly in microelectronics. This has meant a shift in the balance of demand for skilled personnel, away from those trained in traditional engineering disciplines such as mechanical engineering and towards electronic or design and development engineers. Table 1 summarises changes in the distribution of professional science and engineering occupations between 1979 and 1986.

Table 1   Science and engineering professionals, 1979-1986

Occupation	1979	1986
		Percentages
Civil engineering	15	13
Mechanical engineering	13	11
Electronic engineering	10.5	12.5
Design and development	7	16
Engineers and technicians nec	54.5	47.5

Source: Institute for Employment Research, 1988

Science and engineering professionals had the second fastest growth rate during these years, and occupational shifts were mirrored in changes in industrial distribution. Manufacturing industries experienced the lowest growth in the employment of science and engineering professionals (28 per cent) and services the highest (132 per cent). Forecasts to 1995 suggest a continued decline in the relative importance of manufacturing and an increase in the importance of business and other services. Accordingly, the demand for electronics engineers and computer professionals is expected to continue (IER, 1988:24-57).

There are two reasons why changes in the science and engineering occupational mix are likely to provide increased employment opportunities for women. First, the movement away from traditional engineering occupations is likely to dispel the *spanner and grease* image of engineering that has deterred many young women from entering technical employment. Engineering work carried out in an office in front of a computer screen is likely to be more attractive to women and men alike. The second, and perhaps more important, reason concerns the expansion of graduate employment within industry. Historically, young men entering engineering began apprenticeships with a firm at age sixteen, studied for a degree or its equivalent qualification and took examinations for membership of one of the many professional engineering institutions. They followed, in other words, a practical route into engineering and became fully qualified engineers only after several years employment (Smith and Berthoud 1980). Women, for a variety of reasons, did not follow this path into skilled jobs (Cockburn 1983). Consequently, the proportion of qualified women engineers in industry remained very small. In 1979 there were 9,000 women and almost 400,000 men professional scientists and engineers. Although men will continue substantially to outnumber women, it is forecast that by 1995 there will be 68,000 women in these occupational groups. Indeed, between 1979 and 1987, the most rapid increase in the ratio of women to men in any occupational group was in science and engineering professions, although it must be acknowledged that the growth in women's representation was from a very low base (IER, 1988).

While demand for qualified scientists and engineers has increased over the past decade, and appears set to continue increasing, employers have experienced increasing difficulty in recruiting qualified technical people. This has been the case even though the recruitment practices of companies have changed over the last thirty years. The expansion of higher education from the late 1950s and the retention of additional young people in education beyond age sixteen has meant that employers increasingly have recruited graduates in engineering and related disciplines directly from university. However, they have been unable to recruit sufficient numbers of graduates with the skills necessary to realise the full potential of new technology. Few young people choose to study the appropriate subjects to an advanced level while in school and few of those with the relevant A levels go on to choose science or engineering degree courses. There is a high drop-out rate from engineering degree courses and overall, engineering graduates tend to gain lower class degrees. Many new graduates op for occupations outside science and engineering. And an inadequate supply of teachers of technical subjects ensures that the problems of graduate supply are perpetuated.

Methods of research

The principal method of research for the study was an analysis of the policies and practices of ten companies known for their good practice in the employment of women. That is to say, companies were strategically selected for study such that we included only employers who were known to be concerned to increase the representation of professional women staff including qualified scientists and engineers. This approach was adopted on the expectation that *critical cases* would reveal more information in a shorter time than would be available from a random selection of employers. Assistance in compiling our list of employers came from many of the bodies involved in promoting women into science and engineering, such as the Engineering Industry Training Board, the Equal Opportunities Commission, the Women's Engineering Society and the Physics Society. These organisations were able to identify companies who had developed policies and practices to attract, recruit and retain women as scientists and engineers and companies who had contributed to their campaigns and initiatives to attract more women into science and engineering occupations.

There were two stages to the research. In the first stage, interviews were conducted with members of management in each of the ten companies to determine their policies and practices in attracting, recruiting and retaining women qualified scientists and engineers. The number of management interviews varied according to the size of the firm and the number of people with specific responsibilities in the areas of interest to the study. Overall, thirty management interviews were carried out. The interviews took an in-depth form, relying on an *aide memoire* rather than a structured questionnaire. They focused on the nature of the problems which had to be addressed by management if more women (and men) were to be recruited as scientists and engineers, the ways in which these problems had been approached and the achievements of particular initiatives.

The aim of interviewing managers was to get an overview of company policies and the aims behind them. Accordingly, we discussed with managers the development and effectiveness of schools liaison policies and sponsorship schemes. Recruitment managers were asked about graduate recruitment and training practices and the careers which their companies offered to both new and experienced graduates. The issue of staff retention was examined through questions about maternity leave policies, career break schemes and other initiatives aimed at retaining their staff. Interviews with managers were tape-recorded and transcribed verbatim.

The second stage of the research involved interviews with a small number of women and men qualified scientists and engineers in each of the ten companies. (See Appendix 1 for a description of the characteristics of the employees who participated in the study.) These interviews, of which there were 130 in total, also took a qualitative form and were tape-recorded. The aim of interviewing employees was to gain an impression of the extent of knowledge about company policies, to learn how policies were experienced, to discover whether they were thought to work in practice, and to explore the issue of employee support for policies devised, by and large, in personnel departments.

Scientists and engineers were expected to, and ultimately did, express more critical views of company policy and practice than those held by management. Accordingly, measures have been taken to protect their anonymity. Quotations used in the text from our interviews with employees are identified in most instances only by a number and whether the respondent was female or male. Where, however, a company name is given in reference to a particular employee, both number and gender are omitted. We are, of course, very grateful to all employees – and managers – who participated in the study.

The structure of the report

The remainder of the report is divided into two parts. Part 1 begins by examining the ways in which companies have sought to attract young men and women into science and engineering occupations through their schools liaison policies (Chapter 2). Chapter 3 then examines graduate recruitment in the ten companies. Part 2 focuses on retention policies and practices. Chapters 4 and 5 assess employees' opportunities for training and promotion; while policies and practices which companies have developed specifically to retain women scientists and engineers are examined in Chapter 6. The report's findings and conclusions are summarised in Chapter 7.

Introduction

Notes

1. We tend throughout our report to refer to 'qualified' rather than 'professional' scientists and engineers. Qualified scientists and engineers are defined as persons employed in technical work for which the normal qualification is a degree in science or engineering. While the engineering institutions define engineers as 'professional' engineers only when they have acquired chartered status, graduates in engineering are invariably considered as professional staff by their employers.

2. For a report on a conference which addressed the problems of higher education for engineers see *Engineering Futures: New audiences and arrangements for engineering higher education*, The Engineering Council. Among other recommendations, the report suggests that the base of engineering higher education needs to be broadened and more flexible arrangements for entry, transfer and qualification need to be introduced. In addition, conference delegates recommended that campaigns should take more account of women returners in attempting to stimulate demand for engineering higher education.

3. Management interviews were conducted at both the group and business level at ICI, British Aerospace and Plessey.

4. Throughout the report we refer to management interviews and employee interviews to distinguish the two broad groups of company personnel who were interviewed. In practice, however, many of the employees we interviewed had managerial responsibilities, although not in the particular areas of schools liaison, recruitment or retention.

Part 1
Recruiting women into engineering and science

In 1978, women represented two per cent of all professional engineers, scientists and technologists; by 1988, their representation had risen to five per cent. Behind the apparently slow progress of women into the technical field is the fact that fewer girls than boys study mathematics and physics to an advanced level while in school, and that, accordingly, young men continue to outnumber young women on engineering and technology degree courses. The Finniston Report (1980) highlighted what have been perhaps the two most important reasons for the relative scarcity of women in engineering and science occupations: gender differentiation in school curricula and in working patterns.

Boys tend to receive more science instruction in schools than girls. This is so in part because boys are more likely to choose science courses and girls choose arts subjects. Science courses, with the exception of biology, are seen by most pupils as *masculine* in orientation, with the result that girls' self-socialisation and definitions of appropriate gender behaviour discourage them from studying subjects that would lead to careers as scientists or engineers (Blackstone and Weinreich-Haste, 1980; The Royal Society, 1986). Even when girls do choose science subjects, they tend to receive less attention from teachers and spend less time on computers and other equipment, thus leaving uncertain their chances of success and continuing interest in a technical career (Else 1985; Thomas 1985).

Moreover, the requirements of careers in science and engineering, in particular keeping abreast of continually changing and developing technologies, are not readily compatible with women's traditional pattern of labour force participation. It has long been usual in Britain for women's labour force participation to fall sharply from their mid-twenties as family and childcare responsibilities are taken up, and to rise during their early and mid-thirties as domestic responsibilities ease. Consequently, many of the few women who become qualified scientists and engineers are lost from industry as they either fail to return to work at all, or return to occupations which do not make use of their technical skills.

The early 1980s witnessed a heightened awareness of the causes behind the lack of women in science and engineering occupations. In 1984, the Women into Science and Engineering (WISE) campaign was established. Initiated by the Equal Opportunities Commission and the Engineering Council, the campaign enjoyed a high profile and led to the proliferation of other activities at both the national and local level. The WISE campaign, like others such as the Engineering Award Scheme for Women (1980-82), Opening Windows (1987) and Insight (ongoing), was directed at young girls and women and their choice of subjects for advanced study and future careers. Emphasising the attractiveness of science and engineering for young women was seen to be a way of challenging deeply entrenched attitudes about the inappropriateness of technical careers for women.

Monitoring of the numbers of women studying technical subjects and engineering and science degrees suggests that the various campaigns and initiatives have made some impact. Between 1978 and 1985, for example, the number of young women entering A level mathematics increased by 59 per cent. In addition, a survey of entrants into science and engineering degree courses between 1982/3 and 1985/6 found that the proportion of women in engineering had increased from eight per cent to 10 per cent (Engineering Council 1986), while figures for 1987/88 suggest a further increase to 12 per cent (Engineering Council 1988). The Engineering Council estimates that at least 17,000 women chartered and technician engineers will be employed by 2010, compared with 2,700 employed in 1985 (Engineering Council 1985).

Nonetheless, there remains considerable scope for increases in the recruitment of women technologists and scientists to industry. Figure 1 is based upon the findings of a review of employers' recruitment practices in relation to highly qualified people and shows the breakdown of private sector recruitment of women by function.

Overall, women made up 27 per cent of the new graduates recruited to the private sector, but only 17 per cent of those recruited to research and development posts and 18 per cent of those recruited to production and engineering.

Figure 1 Private sector recruitment of women new graduates by function

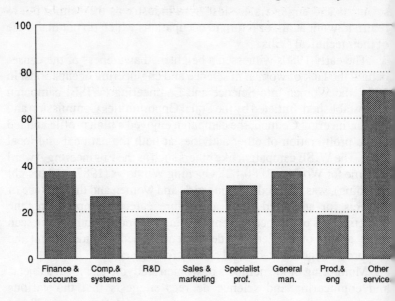

Source: PSI and IER 1990

In the next two chapters we examine corporate policies and practices in relation to the recruitment of women. Chapter 2 focuses upon schools liaison policies. Companies become involved in the education of pupils in order to improve the supply of labour to industry by encouraging young people to enter technical careers. Companies are concerned to influence the educational choices of boys and girls alike. But in recognition of a gender imbalance which favours boys, both in the study of technical subjects and, ultimately, in the decision to enter technical occupations, companies have increasingly directed their efforts towards encouraging girls to choose subjects that leave open the possibility of a technical career. Accordingly, our concern in Chapter 2 is to describe and assess the strategies undertaken by companies in relation to the educational choices of young people in general, and the choices of young women in particular.

Chapter 3 focuses upon graduate recruitment. The ten companies we studied were medium and large recruiters of new graduates. In general, company policy and much practice was favourable towards the recruitment of women. Managers had responded positively to the increasing number of women graduates and were anxious to recruit women into technical jobs.

Note

1. See Peter Elias and Malcolm Rigg (eds), *The Demand for Graduates*, Policy Studies Institute and Institute for Employment Research, 1990.

2 Schools liaison

Schools liaison activities accelerated during the late 1970s and 1980s, when a wide variety of projects were initiated to improve links between schools and employers. In 1978, for example, the Department of Industry established an Industry/Education Unit which provided teaching materials designed to attract children into science, engineering and technology. In 1986, during Industry Year, the CBI financed an 'Understanding British Industry' initiative and organised a number of related events, competitions, and industry study tours. A report by its education task force called for the establishment of a national body which would increase the number and quality of industrial links with education. At the same time, the Engineering Council began to promote a Neighbourhood Engineers project, with the aim of linking 24,000 engineers to 6,000 secondary schools across the country.

The surge of interest in links between employers and schools led to a substantial increase in requests from schools to companies for resources, work placements, careers advice and other forms of involvement. Managers reported that initially their companies had tried to satisfy all requests, but the sheer volume of demand ultimately made it impossible for them to do so. There was a gradual realisation that companies were reacting to external demands rather than deciding their own aims and priorities. This led to the appointment of managers with the specific task of overseeing educational policy. One manager at Esso explained:

> About two and a half years ago, I was appointed because the department manager at the time felt we had rather lost our way. The chairman of the board felt we had lost our way in terms of

what we were doing and what we should be doing in our contact with schools.

The demand from schools for companies to provide teaching resources, attend careers conventions and provide other forms of assistance also pushed managers into considering the most effective ways of using their resources:

> Once one is a big company which is known to all, one tends to get leapt on by everyone. So we are trying to specify a) what we are trying to achieve and b) some priority objectives and c) some methods which are seen to be cost-effective.

Our concern in the present chapter is to describe and assess the strategies undertaken by companies in relation to changing the educational choices of young people in general, and the choices of school girls in particular. The range of activities undertaken by the companies in schools is discussed, with particular attention paid to the ways in which companies have tried to attract women into the technical field. We also assess the influence of teachers and parents and explore the response of pupils towards schools-industry links. The chapter ends with an examination of the extent to which managers have been able to evaluate the effects of company policy and their future plans for schools liaison.

Putting policy into practice

The activity of creating and sustaining links between employers and schools was undertaken primarily at the local level, although all of the companies had managers in their central or head offices with specific responsibility for schools liaison policies. The number of links with schools varied according to the size of the company and the number of different sites at which branches were located across the country. Westland, for example, had long and well-established links with a small number of schools at its major site in Yeovil. ICI and Esso, in contrast, had extensive networks of links between their numerous sites and local schools throughout Britain. Rather than trying to be *everything to everybody*, the Esso schools liaison manager had established a system of *link schools* so that most Esso establishments in the country were linked to a small number of schools nearby. He explained:

> We have actually got fifty link schools now up and down the country and what we are trying to do is build up an in-depth

relationship with those schools such that we can put over the science and engineering messages best into those schools, in-depth over a number of years, with both the teachers and the students, so that we can really effect a change in that relatively small number of schools.

Proximity was important so that links between the site and the school could be easily sustained and close ties developed over a number of years. Both the company and the school could build upon a range of activities in ways that would allow a cumulative effect. However, managers were also concerned about creating links with certain types of schools. In all cases, there was a strong desire for companies to attach themselves to good academic schools. In addition, managers wanted to ensure that they developed links with single-sex as well as mixed schools. Establishing links with girls' schools, in particular, was seen as a way of demonstrating their wish to recruit women as well as men, and of establishing their broad commitment to equal opportunities.

The companies were engaged in a range of activities in the classroom. Managers argued that technical subjects were not perceived as exciting or engaging by most young people. They aimed, therefore, to present positive images of science and technology, dispelling the myth that subjects like physics and chemistry were uninteresting and merely required the rote learning of facts and figures. Marconi Defence Systems, Stanmore,* which was twinned with a local girls school, provided activity-based learning sessions for each year in the school. Activities included the design of simple electronic goods: pupils had to take account of the production schedule and to cost the product in accordance with design specifications, giving them an insight into the engineering and the business side of industry. Marconi also sponsored *a great egg race*, in which managers and teachers acted as advisers and judges of the girls' engineering skills. Employees from Plessey Controls in Poole ran eight-week courses in two local schools teaching basic electronics. Westland had also devised school projects and activity packages. In general, companies designed projects and activities which were intended to give young people some understanding of technical subjects, but ones which, at

---

*   *Marconi, Stanmore, is part of the GEC group.*

the same time, were exciting and which maintained *all the child-like interests, the imaginative and creative things.*

Companies were particularly concerned to use up-to-date technology in the classroom. The practical applications of new technology were promoted through company visits and work placements. Employers hoped to communicate the importance of studying technical subjects at school by showing pupils the relevance of technology to many aspects of people's lives. The schools liaison manager at Ove Arup explained:

> They want to see people using maths. Some people want to see them use computers. Others will want to see people at drawing boards. We do all that and we show them a film of one of our projects from inception to completion. These are very stimulating so the teacher can say 'look – this is the subject I teach you. This is its application in the world outside. This is how it is used. This is why you need the subject'.

Work placements were a way of showing young people what engineers did on a day-to-day basis and the industrial environment in which they worked. They also could aid recruitment for the companies involved, by encouraging young people to seek sponsorship and employment at a later stage. The managers we spoke to were paying increasing attention to the quality of work experience provided on placements and aimed to provide pupils with genuine training rather than low-level technical work. The manager at British Aerospace reported:

> One of our priorities at the moment is improving the quality of work placements. It is not just a case of providing so many placements for a school but also actually thinking more about what will be achieved from the work placement, what the school is trying to achieve from the work placement and then developing quality placements which we would monitor and review.

All of the companies participated in careers talks at their linked schools as well as taking part in national careers fairs and conventions. They aimed to provide young people with positive images of technical jobs, portraying them as demanding, professional, highly paid and as having good opportunities for career development. The manager at STC explained:

> One of the messages that we are trying to get across is that engineering is about solving intellectual problems and it is done by people in white-collar jobs in sophisticated offices. They are

> solving intellectual problems at a very high level. It is not done with spanners and grease and oil. Anyone who sees engineering in terms of spanners does not know what they are talking about.

Dispelling the *spanner and grease* image of engineering was seen as particularly important in order to attract women into the company. Managers made every effort to include women scientists and engineers as speakers at careers talks and conventions:

> It is all right sending a man and saying 'oh yes, girls can do this job as well', but there is nothing like sending a women there to say 'I do this job and I do not walk around in a boiler suit and boots and a crash helmet (Schools liaison manager, Ove Arup).

Some of the larger companies such as Esso, ICI and Boots emphasised to pupils that they had a broad range of occupations within their organisations. They pointed out that technical subjects were a good springboard to any career, not just one in engineering or science. Echoing the views of other managers, the schools liaison manager at Esso suggested:

> We are trying to put over to the kids the idea of a scientific or technical education leading to the general job market because it is easier to do it that way round than, say, taking a humanities degree and then trying to grasp the engineering side afterwards.

Smaller companies also were anxious to show that they offered the prospect of promotion into sales, marketing or general management, for which background technical knowledge was necessary.

Any policy has to be implemented by the individual employees who make up an organisation, and schools liaison was no exception. Many of the employees we interviewed were actively involved in their companies' schools liaison programmes and took part in showing young people around their workplaces, visiting schools to oversee classroom projects, and attending careers fairs. Managers acknowledged that, to be effective, school liaison policies relied on the dedication and goodwill of their employees. They liked to use young enthusiastic engineers and scientists to present projects in the classroom. Younger employees were expected to convey to pupils their own interest in and enthusiasm for technical activities. However, personal contacts between school pupils and older engineers also were seen as valuable:

> It is also important to have people who have been in engineering a lifetime and still have a great warmth and humanity about them and a cheerfulness and they still enjoy coming to work (Schools liaison manager, Ove Arup).

Ove Arup gave their employees special training for speaking and working in schools, building up their self-confidence and encouraging them to accept invitations from schools to work alongside young people. Indeed, managers acknowledged that contact between company employees and young people benefited individual employees as well as the organisation as a whole:

> Our engineers that went into the schools gained a lot of benefit in developing their effective communication skills. They had to prepare the lesson material. They had a significant amount of planning, problem-solving, and decision-making to undertake. They had to liaise with members of the teaching staff as well as actually deliver the material and keep the attention and reasonable discipline of a group of 11 year olds – a difficult population to handle (Schools liaison manager, Marconi Defence, Stanmore).

Encouraging young women into engineering and science

Company managers were well aware that girls are less likely than boys to study technical subjects and that many girls need special encouragement if they are to choose subjects which are traditionally seen as unfeminine and inappropriate. Girls who chose to study technical subjects would often need to justify their unusual decision to family, friends and even teachers. For the male employees we interviewed, choosing to study science subjects was uncontroversial, even expected:

> I think I was influenced by the fact that men did sciences quite strongly and it was generally the feeling in the school that most of the lads did science subjects and very few did arts.

Many men said that studying science followed naturally from their personal enjoyment of dabbling in mechanics and electronics, and from playing with computers. Women employees were similar to the men in that science subjects had generally been their best subjects at school and those which they had most enjoyed, but they were more likely to mention the personal support of a parent, teacher or particular friend as an important factor encouraging them to pursue a technical career.

Managers at Westland had discussed the issue of subject choice with their women employees in order to gain some insight into the problems which had to be tackled:

> We looked at the reasons why by talking to the girls who have come through. They have all shown common factors. They all found that they had passive resistance either from parents, from school and from careers services. However, they all had one parent in industry or technology. It could be a very humble level but they all had some sort of link. They all found that they were on their own at school wishing to become engineers. They've had to stick to their guns. They have had to be of quite dogged spirit to see their way through to the end.

Companies were particularly interested in persuading girls to take an interest in technical subjects and, for this reason, they gave special encouragement to women employees to participate in liaison work with schools. A woman engineer from Ove Arup recounted that

> ...they encourage everyone but they push their women forward much more to go out and talk to schools.

Since women scientists and engineers were a minority in the companies we studied, the burden of schools liaison work fell disproportionately on their shoulders. However, the women employees we spoke to did not, in general, mind being singled out for these activities. They found contact with schools enjoyable, and stressed the importance of female role models. They emphasised the importance of giving young people an honest and accurate impression of the opportunities open to them, drawing on their own experiences, career choices and decisions, and describing aspects of their jobs which they enjoyed as well as the parts which they disliked.

Managers gave considerable attention to the portrayal of young women in the materials, such as videos, magazines and company literature, which they provided to schools. Both ICI and British Gas had extensive catalogues of materials which they provided to schools at cost. They sought to avoid stereotyped images of women and to provide positive images of successful women in the technical field. Marconi Defence, Stanmore, was involved in redesigning its company literature in liaison with the girls school with which it was linked, and was relying on the literature to present a favourable image of technology to school girls.

Managers recognised that science subjects often were not taught in a manner that might attract young women. The manager responsible for schools liaison at ICI acknowledged that:

> There is a great deal of evidence that the way physics is taught – the instances of applications are often military. They are often aggressive. They are about motor cars and they are about missiles and those sorts of things. They are of very little interest to girls. They could be about scanners, they could be about hospitals, they could be about children.

Using the findings of the Girls Into Science and Technology (GIST) project (Whyte, 1986), managers at British Gas were producing more *girl-friendly* material for their catalogues of videos and company literature. The schools liaison manager explained:

> We have in mind a technology film about ... developing a gas fire or something like that rather than developing a testing technique for an off-shore structure. The evidence is that things which are more closely related to human benefit seem to be more motivating for the girls in science and technology, so we bear this in mind.

Managers stressed that it was important to work with educationalists and others *to reinforce efforts made to change the gender bias in science and engineering*. They argued that the ways in which technical subjects were taught in the classroom needed to be changed if young girls were to be attracted to study them. Ultimately, however, teachers and educationalists, rather than companies, had more say over these issues, and managers tended to recognise the limitations on their own influence.

The role of teachers

Teachers obviously have a very important role to play in shaping young peoples' decisions about their futures. One manager reported:

> We are very keen on the idea of influencing change by influencing the teachers. We then get what we call the multiplying effect coming through. If I can convince one teacher that engineering is a good subject the chances are that they in turn will convince generations of school kids that engineering is a good subject.

The companies in the study encouraged visits to the workplace by teachers and also operated work-shadowing schemes, whereby teachers could gain awareness of industrial jobs, insight into the applications of new technology and ideas for project work in the classroom. Managers felt that it was important to dispel misconceptions of industry, even among science teachers, since few of them had any industrial experience. Work-shadowing was also a way of challenging the attitudes of teachers who were opposed to women entering technical occupations. Such attitudes had been found to exist and were, according to one manager, the result of *years and years of distillation that says these particular kids go to this particular job.*

However, schemes involving teachers were still in their infancy. The company with the most advanced policy, Esso, had provided industrial experience for only 25 teachers. A woman engineer from British Aerospace commented that:

> One of the things which I feel is missing very drastically is the fact that engineers don't have any communication with the teachers that teach children. Teachers often have preconceived notions about what particular jobs are and why they wouldn't like to do them which are passed on. Kids are very much influenced by those people, so getting teachers into industry and experiencing what it's like so they can physically describe what happens to the kids to make it more realistic is important.

Managers were concerned also that they mainly met those teachers who already had positive attitudes towards technology. A manager from Ove Arup remarked:

> I think teachers are very good but then I suppose I meet the ones who come to us. I meet the positive teachers rather than the negative ones.

Managers felt that closer liaison with teachers could be fruitful and they intended to devote more resources to it in the future. In the view of most managers, it was in their self-interest to do so, and teachers, who faced staff shortages in key subjects, restricted in-service training and limited resources, needed all the help that companies could provide.

Pupils' responses to schools liaison

In general, managers responsible for schools liaison felt they had received a favourable response from pupils. Young people enjoyed the project work, visits to sites and work-shadowing and participated enthusiastically in the range of activities organised by the companies. Boys and girls alike were inquisitive and interested in what was being asked of them. They enjoyed the new situations in which they found themselves and liked doing project work which differed from their usual school work. The manager at Marconi Defence, Stanmore, noted that the girls at their linked school enjoyed *the novelty of a new face in the teaching situation* and reacted positively to the *improved ratio of teachers to pupils*.

The most positive feedback came from sixth form pupils, some of whom had attended courses such as the EITB's *Insight into Engineering*, to which many of the companies contributed. There was a very positive response from young women as a result of hands-on experience of engineering techniques. It was felt, however, that these pupils were a self-selected group, already favourably predisposed towards the sciences and technology. Where the companies may have been successful is in persuading them to take an applied science degree, such as chemical engineering, rather than read the pure sciences. Managers acknowledged that they seemed to be more successful in preaching to the converted than in reaching the larger mass of young people.

Employees who had been involved in work with schools also found that young people tended to participate enthusiastically in project work and listened to careers talks with interest. Nonetheless, employees reported that the response from young people was disappointing on two counts. First, employees felt that young people had limited preoccupations. Their questions tended to focus on what combination of A level subjects they should take or what grades were required for entry into university. According to our employees, boys and girls alike were concerned more about their immediate studies than in the range of careers which they might pursue later.

Secondly, employees found that negative stereotypes about careers in industry were still rife. Engineering, in particular, was considered to be a manual occupation predominantly undertaken by men. As a female engineer at Marconi Defence, Stanmore, explained:

> I remember starting out a talk recently on what do you think an engineer is, and being both shocked and surprised when they came up with the archetypical sort of man in a boiler suit in an engine room. I would have liked to have felt that things had changed over the past few years but it seems not to have done, not for everybody anyway.

Employees noted that young women continued to be concerned about working in a predominantly male environment and, more specifically, about problems of sexual harassment and discrimination. While they responded positively to female role models, these issues still came to the fore. A female engineer at Esso reflected:

> They like seeing women who have done well and they ask all the typical questions: 'Do you have problems with things like discrimination?'

While talks were well received, employees found that only a small number of young people genuinely were interested in following technical careers. Some employees expressed regret at being unable to see any concrete results of their activities in schools; one employee suggested that pupils *might enjoy an exercise and find it all great fun, but then quite what you're left with at the end of the day is quite hard to say*. It appeared to many employees that a small self-selected group would continue to pursue technical careers, but that the career choices of most young people were not changing as a result of company policy. Accordingly, employees stressed the importance of talking to children of primary school age. If very young children could be persuaded to develop an interest in technology before stereotyped attitudes towards technical subjects became firmly entrenched, it was argued that they would be less likely to drop their science options at the first available opportunity.

Moreover, the response of young women at careers talks and fairs suggested that many dismissed the idea of studying technical subjects at university and pursuing a technical career in industry. They preferred jobs deemed more appropriate for their sex, with only a small number of women appearing willing to consider technical occupations. The manager at Ove Arup said:

> I think a lot of them have already made up their minds. Even before they take their options, I think some of them have already blotted out certain areas. I do not know if it is home influence, school influence, media influence but they seem to have blotted out certain things already. They are not receptive.

Enjoyment of technical projects at school had not, it seemed, led young women to choose technical careers. Our findings reflect those of a study on the effects of a roadshow encouraging women to enter the technical field (Pilcher et al 1986). Pilcher and her colleagues found that the young women responded positively to the event and were open to new ideas. But while they reacted positively to careers advice at the time, their enthusiasm did not translate into action after the event. It seemed likely from this study, as from our own, that peer or parental pressures counteracted the positive views generated by industry-school links in ways that meant the young women did not choose technical careers at a later date.

Nonetheless, all of the employees who had been involved with schools liaison projects felt that they were valuable. They enjoyed the opportunity to talk to young people and to relay information about careers in science and engineering from which young people could make informed career choices. The provision of up-to-date information on opportunities in science and engineering was particularly important for young women if they were to overcome stereotypes and prejudices about women in technical jobs. A female engineer at Esso explained:

> It is important to prove to girls in school today that it's not just something you read about. I will tell you that I've done it and prove that it isn't as difficult as you think it is.

Company evaluation of schools liaison initiatives

Few of the companies had evaluated their schools liaison policies. Most of the policies had been in place formally for only a short time and managers felt it was too early to measure any changes in attitudes or educational and occupational choices. Managers also recognised that problems were likely to arise in relation to evaluations based upon quantifiable indicators of success. Accordingly, managers tended to discuss the evaluation of different components of schools liaison policy rather than the policy in general.

Managers at Boots and British Aerospace, for example, spoke of ways of evaluating work placements and suggested that the benefits of work placements should be reviewed at the completion of work experience and the results monitored. A manager at Boots explained:

> At the moment, I'm finding it extremely difficult to establish the benefits people have gained from each of the programmes. I've tried to establish for everyone what the aims of a particular involvement are and then at the end, try and review what we set out to do and ask 'have they been done?' Then I am able to judge whether the aims have been achieved.

It is widely accepted that educational change leads to wider social change only very gradually. Moreover, the measurement of such changes that do result from the actions of companies in schools is particularly difficult. It would not be clear, for example, whether the increased number of qualified women entering technical occupations was the result of schools liaison policies or whether it was a by-product of the growing number of women attending university. The causal links between schools liaison policies and increases in corporate graduate intake of women can not simply be inferred, and would need to be explored in more depth than has been possible in this wide-ranging study.

Our evidence suggests, however, that to date schools liaison projects appear to act as a positive but limited influence on the educational choices of some pupils. The limits to the effectiveness of links with schools were found to be both quantitative, in that companies were able to reach only small numbers of pupils and teachers, and substantive, in the sense that too often the pupils and teachers reached were a self-selected group, already receptive to the idea of technical careers in industry. The limits on the effectiveness of schools liaison projects were recognised by the managers and the employees with whom we spoke. Nonetheless, there was widespread agreement that such projects did have beneficial effects, albeit ones that were small and, as yet, unmeasured.

The future of schools liaison

All of the managers we spoke with foresaw increased involvement with schools in the future. They would continue to develop closer links with local schools and with schools across the country, and accepted that results from these links would become evident only in the long run.

Closer links with primary schools were considered to be a major priority for the future. It was felt to be important to try to instil positive attitudes towards technology in young children before *social conditioning* had time to leave its mark. The manager at Esso stated:

> It is certainly an area I want to get involved in because I think that is where some of the attitudes are made, particularly for females. At primary level, the girls go to that side of the classroom and the boys go to that side and the girls do cooking and the boys do whatever they do. I feel there is a general groundswell saying this is wrong.

Companies would continue to contribute to, and participate in, national schemes such as the Engineering Council's *Open Window* and *Community Engineers* schemes. They would continue to support and liaise with organisations specifically concerned with attracting women into the technical field such as the Women's Engineering Society and the Physics Society. Coordinating activities with other organisations who already had developed expertise in dealing with schools and who knew *where you should apply your shoulder and push if you want to be most helpful* was seen as a priority for future policy. Companies' actions would have greater effect if carried out in tandem with other organisations with the same aims and objectives.

But despite the importance attached to schools liaison policies, managers reported that their activities could be curtailed by limited resources. Schools liaison policies had relatively small budget allocations. One manager compared his department with *a rowing boat alongside the Queen Mary*. Schools liaison was susceptible to budget cuts because it did not produce immediate, quantifiable results. A manager from one of the smaller companies of the study explained:

> I do not think we do anywhere near enough to make any form of impact. The reason for that is we do not have the investment. For example, my budget was cut by £80,000 this year. Every time it happens I take the schools programme out. I must get sponsored students. I must get graduates. I have got to set a priority and in my set of priorities schools liaison is the lowest one.

Moreover, financial resources are not the only ones threatened. The manager at Westland suggested:

> The limitation comes in the [human] resources. Whether a training board contacts us or just a school rings up and asks if someone can speak on such and such, the answer is yes we would love to do this but the number of people we can draw upon to give the lecture is the same however many different organisations are involved in asking us.

The problem of arranging women speakers was particularly acute, as there are not enough women in technical jobs in industry to provide the role models that are urgently needed. Finally, the success of schools liaison policies, managers argued, depended on a well-resourced education system. A teaching profession which enjoys better status and higher financial rewards was needed so that those with technical qualifications will be attracted into teaching. Otherwise, teacher shortages in key subjects like physics and mathematics would only exacerbate existing labour supply problems.

3 Graduate recruitment

In this chapter we focus on the graduate recruitment policies and practices of the ten companies under review. Throughout, we pay particular attention to the recruitment of women. We consider issues concerning equal opportunities policies, positive discrimination and individual prejudice against the recruitment of women and look at the provisions made for training for recruiters in the different companies.

Many of the employees we interviewed had experienced their companies' recruitment procedures from both sides: as recruits and as recruiters. Accordingly, we make use throughout the chapter of their experiences in both of these ways. Older employees were more likely to base their views on their experiences as recruiters. Their memories of being recruited were more distant than those of their younger colleagues, and were usually less relevant to the current situation. Younger employees were unlikely to have been involved as recruiters in all stages of the recruitment process, but were able frequently to recount their recollections of being recruited.

Recruiting graduates

The ten companies we studied were medium and large recruiters of new graduates. Recruitment on this scale involves a large influx of new employees into the company at a single point in time and, accordingly, requires a considerable administrative process. Moreover, because of increasing competition for technically qualified graduates, all of the recruitment managers reported that graduate recruitment was becoming a year-long activity.

The first step in the graduate recruitment process is an internal one, as recruitment managers establish the demand for scientists and engineers within their organisation. All of the companies we studied followed similar procedures for calculating demand. Those responsible for recruitment usually contacted the different parts of the company, requested the number and type of vacancies that needed to be filled and compiled an overall figure of graduate demand. This final figure would subsequently be ratified at board level. Recruitment managers would then publicise to students the types of vacancy available within their organisation and their numbers.

Company literature

The graduate brochure is one of the major ways in which companies advertise the number of vacancies on offer and the degree disciplines in which the various parts of the company are interested. The graduate brochure is also a means by which companies attempt to *sell* themselves to students and encourage final year students to make job applications.

All of the companies we studied emphasised a broadly similar range of benefits which they considered to be particularly attractive: good rates of pay, the possibility of travel and work abroad, training programmes, personal development plans, interesting projects and promotional opportunities both inside and outside the technical field. The larger companies of the study such as Esso, ICI, British Gas, British Aerospace and Boots highlighted the varied career opportunities which they could offer new graduates. Companies which were not solely engineering environments, with the possible exception of British Aerospace, employed both science and arts graduates in some departments, for example, marketing, planning, supply and distribution. For the most part, these companies were seeking potential managers, not technical specialists. The recruitment manager at Esso explained:

> We need a few technical experts but we do not need a lot, and more and more we are recognising that you can buy that for a week from a professor or an agency or what have you. We want generalist managers.

The predominately engineering companies, like Westland, Plessey, Marconi Defence Systems (Stanmore), STC Telecommunications and, to some extent, Ove Arup, emphasised the challenging and interesting technical work which a new graduate could expect to enjoy on a variety of different projects. At Westland, for example, the company's own research had found this issue to be of concern to students. In attracting graduates, Westland stressed that it was

> ...a company involved in high technology because most graduates on the engineering side would like to feel that they are doing something of major technical interest.

The graduate brochure was also used by the companies to attract women undergraduates studying technical subjects. The companies used similar methods to show their interest in employing women as scientists and engineers but some companies gave a higher profile to their equal opportunities policies than did others. Ove Arup, Esso, British Gas, STC and British Aerospace all made explicit reference to their commitment to equal opportunities. Each emphasised that recruitment and promotion within their organisation would be determined by individual merit as defined by job performance. In ICI's graduate brochure, for example, the company emphasised that:

> Our stated policy and established practice require that recruitment into the company and progression within it will be determined solely by personal merit and criteria relating to the effective performance of the job and the needs of the business.

Not surprisingly, companies who had been successful in recruitment of women as scientists and engineers choose to advertise their achievements. Ove Arup, a construction engineering consultancy, stated in its brochure that it employed the largest number of women engineers in comparison with any other engineering consultancy in the construction industry. The Esso 1989 graduate brochure put a figure to its claims, noting that 39 per cent of its graduate recruits in 1988 were women. ICI, like Esso, pointed out that one in three of their graduates recruited from university in 1988 were women and noted that the number of women in senior positions was also gradually increasing.

A general feature of graduate brochures is the *employee profile*, a brief account of the career of a successful employee, which serves to illustrate the opportunities that exist with the company. Almost all of the companies used women scientists and engineers in their employee profiles to emphasise that they were keen to employ women and had done so in the past. Recruitment literature portrayed women engaged in technical activities and managerial jobs. A recruitment manager at British Gas reported that:

> We consciously try to get pictures of women engineers in the publicity material. It is a conscious effort to convey that we are interested in women.

Recruitment managers felt that it was important to show that women were already employed as scientists and engineers. While recognising that employee profiles gave an overstated impression of the number of women employees in the technical field, they believed that it was one of the few ways available of attracting women into the company. A manager at Plessey, a company which includes a number of women employee profiles in its graduate brochure, reported his reasons for including women in the brochure:

> I would be very dismayed and so would my senior colleagues if we didn't have women figuring as both scientists and engineers because there is nothing like a comfort factor of a women undergraduate seeing somebody who looks rather like her already working for us. So we're reinforcing the view that... we can offer you a career and there are already people like you in there who have careers. So it's comfortable and warm, come and join us.

A number of companies advertised the provisions they made for maternity leave. Ove Arup, for example, emphasised the way in which they recognised the special needs of female employees by providing *maternity provisions which are both generous and flexible enough to suit each individual case*. Esso literature drew attention to the company's career break scheme, providing details and emphasising that it *is open to everyone, men and women*. Other companies, such as Marconi Defence Systems (Stanmore) and ICI, both of whom had recently devised career break schemes, planned to incorporate information on company maternity policy in future recruitment literature.

Finally, a number of companies referred to their involvement in national initiatives aimed at attracting more women into science and engineering occupations. For example, Ove Arup and British Gas made reference to their support for and involvement in the WISE campaign, stressing their interest in attracting women into technical occupations.

Talks and presentations

Because of the high degree of competition for graduates, the companies were increasingly engaged in making presentations and giving talks, attending careers fairs and conventions, running courses about the company and establishing university contacts with relevant departments as early as possible in the academic year. Managers felt that it was important to establish their presence at universities and polytechnics, making their company name known to students. In turn, it was hoped that final year undergraduates would be attracted to their company before other companies, and especially city accountancy firms, made known their own graduate opportunities.

According to recruitment managers, the overriding benefit of talks, presentations, courses and other similar activities was the personal touch. Companies recognised the students' desire to go beyond the glossy brochures, and some expressed the wish to give students a balanced view of the company, including its good and bad points. The recruitment manager at ICI explained:

> I think all the research shows you can advertise, you can put entries in all of the directories, you can print brochures, all of which are very important, but the thing which students are very interested in is actually meeting people from different companies. So if you are sending along inspiring young managers of both sexes to meet students and to talk to them and give them the opportunity of finding out, that is going to encourage them to join us.

Close attention was paid to the selection of staff to represent the company at talks and presentations. All of the companies were keen to use their previous year's intake of new graduates to promote the opportunities available for new recruits. However, the larger companies like ICI and Esso were generally the most explicit in their use of relatively new employees who, as young people with recent experience of being recruited into the company, were best suited to talk to potential recruits about their hopes and plans, fears and misgivings. Moreover, as the recipients of good rates of pay, varied opportunities and an active training programme – at least in the early years of employment – the previous year's recruits were expected by their employers to be inclined to promote the company in a particularly good light.

Management made a conscious effort to include women representatives as part of this process. Again, one of their aims was to show that they employed women who enjoyed their work, and that some women already occupied managerial positions. As one recruitment officer suggested:

> We make maximum use of every women that we employ, if we possibly can, to get them to go and give talks to schools [and] to universities at career days.

Companies, of course, must employ women in the first place in order to be able to present women scientists and engineers to the student population. Managers argued that when women worked within the company in larger numbers (or in more equal proportions to men), it would no longer be necessary deliberately to choose women to represent the company.

Company sponsorship

One in four final year engineering students in the universities and polytechnics are sponsored by industry. Sponsorship does not appear to be used as a special means to attract more women into engineering and science careers, although there is evidence which suggests that women are increasingly likely to apply for sponsorship and are slightly more successful than men in obtaining sponsorship (Connor and Gordon, 1985). From the perspective of the firm, sponsorship schemes ensure a supply of engineers, and may be one way for industry to help to increase the future supply of graduates. All of the companies we studied operated sponsorship schemes. Indeed, as a form of long-term recruitment, managers foresaw an increase in the number of company sponsorships they offered.

A large number of applications for company sponsorship were received each year by the companies we studied. Marconi Defence, Stanmore, for example, received more than 500 applications for 25 sponsorships. The overwhelming majority of all applications were made by men. Managers reported that they were disappointed by the lack of number of women applicants because it suggested that women were continuing not to choose technical careers in industry. The manager at Plessey Controls in Poole reported:

> We had a very low response this year. We disseminated all of our local information and publicity throughout the local schools and colleges but the response from young women has been very, very low this year. We offer apprenticeships every year. This year was exclusive. Just the boys applied for it.

In the companies we studied, then, there was no evidence of the general upward trend in the number of women embarking upon technical careers through the sponsorship route, as noted above.

Managers felt that there was little they could do if women did not apply to the company for financial assistance during their university careers. Targeting of sponsorship publicity towards young women in local schools had not been considered as a way of increasing the number of women who came forward for sponsorship.

Attracting women graduates: the employee view
The employees we interviewed were asked a number of questions about their companies' recruitment policies and procedures. A key question was whether they thought that their company was genuinely interested in recruiting more women scientists or engineers. Unsurprisingly perhaps, the question produced a wide range of responses, which reflected the experiences of each individual as much as the policies of particular companies.

Overall, employees recognised that within the standard graduate recruitment process their companies had devised various means of showing interest in employing women and of drawing attention to their equal opportunities policies. Many employees suggested that an equal opportunities image was advantageous for companies which were thus able to portray themselves as not only fair but also as modern and progressive in their employment policies. However, a few employees felt that the emphasis on the employment of women was little more than a public relations exercise:

> It's almost like paying lip service and they're quite keen to be seen employing women (069, female).

> I don't know how much of it is actually PR. You know, 'it's good for us to be seen doing this' (044, female).

> It's put *we are an equal opportunities employer* on all its advertisements, for example, so I think the words are there, but perhaps the action isn't yet (012, female).

Moreover, some employees argued that, as competition for female recruits becomes more intense, companies would have to do more than show that they are prepared to employ women on the terms that currently exist. Statements of company policy are a first step it was argued, but they need to be followed by action if they are to be credible to recruits and employees alike. It was suggested that companies could do more to promote the advantages of engineering as a career for women:

> There could be a bigger push by the company, emphasising that you can have female engineers and that you don't have to get your hands dirty and there are good chances for promotion and it is an exciting job, and the pay is comparatively a lot better than traditional female type jobs (069, female).

In addition, employers need to place more emphasis on the practical benefits which they can offer women: the improvement of working conditions for parents, through the provision of creches, job-sharing, part-time working and other initiatives, was mentioned by several employees as an important first step.

Equal opportunities policy

All of the companies had had equal opportunities policies in place since the mid-1980s. In accordance with these policies, managers reported that their companies were committed to employing women as scientists or engineers and to ensuring that women enjoyed equality of opportunity for training and promotion. Recruitment managers identified a number of factors which had contributed to a sense of *openness* towards employing women in the technical field, and to an ethos of equal opportunities.

First, companies had taken advantage of the growing number of women leaving higher education with degrees. Women graduates have made a significant investment in education and training, and tend to have long-term career aspirations. Included among this group of women are the small but increasing number of women who take degrees in technical subjects. As the recruitment manager of Esso reflected:

> I suppose that it must have been about thirteen or fourteen years ago that we first saw women appearing in a sort of professional role. Personnel probably had one or two but it was almost unheard of [anywhere else]. There was the odd token women... Now I would say 100 per cent across the company, we have reached the stage where we do not even talk about it anymore. It is a fact of life. People go and look for four graduates. Maybe they are all women. Maybe there is one woman. I suppose we would be quite surprised if they had four graduates and none of them was a woman but if they were all women, it would not surprise us now.

Secondly, there have been changes in the traditional routes of entry to the technical professions and to the engineering profession in particular. As Berthoud and Smith (1980) found in the late 1970, engineers increasingly take an academic route into professional engineering, acquiring their credentials though the education system, rather than acquiring training, skills and qualifications through a practical route within the firm. Managers noted that these changes in the engineering profession have allowed women an easier entry into professional engineering.

Thirdly, managers described how their organisations had responded to a social climate where equal opportunities had become a salient issue. Awareness of the small number of women found in technical careers in particular had been heightened by a series of campaigns and initiatives organised by the Engineering Industry Training Board, the Engineering Council, the Equal Opportunities Commission and others since the mid 1970s. The most prominent of these initiatives was Women into Science and Engineering (WISE) 84, a year-long campaign to attract young women into the technical field. One successful outcome of these campaigns was, it seems, to raise the level of management interest in the issue:

> I suppose I would suggest that it was the WISE year that really suddenly raised it as a major issue, albeit Plessey, through the then Director of Personnel had been aware of the problem for some time. He did raise it within the circles within the company at a higher level than had been envisaged at that point in time. The WISE year put a cap on top of that in terms of 'here are women as a particular issue, let's try and encourage them in some way or other to begin to think about engineering as a career'.

Finally, and perhaps most importantly, managers acknowledged that persistent problems of supply have meant that companies simply cannot afford to disregard the growing number of women with technical degrees. As a manager at Plessey stressed:

> We cannot afford to turn anybody down, male or female. It does not matter. We want engineers. We're not looking at males and females. We are looking at people, engineering people, and there are not enough of them.

Although most employees acknowledged that their companies genuinely were interested in recruiting women scientists and engineers, many of them attributed negative reasons to this interest. For example, employees pointed out that companies could no longer choose to exclude women; legislation and their own demands for labour required them to consider women applicants. The dwindling supply of traditional male recruits, the problems of skill shortages in engineering and changes in demographic trends were frequently cited reasons for employers' attempts to recruit women. Companies were described as understaffed and as having recruitment difficulties. Many employees said that their companies were interested in recruiting *full stop* – they did not have the luxury of discriminating between men and women.

> They would choose men if they had a choice but they don't have a choice, therefore there's no discrimination (038, female).

> From the company's point of view, if they can get a man to do the same job and not have to pay out for looking after the children... then it's in the company's interest to employ the man on that basis. But it they're short of staff and women with children are the only ones available that can do the job then I suppose they'll have to (007, female).

> Some of the other female graduates... got the impression that some of the people interviewing them wouldn't employ a woman if they had a choice in the matter (068, female).

A small number of employees expressed the view that women were seen as being cheaper to employ than men. There was a feeling that women were more committed to their families than their careers, and this had some benefit: *if they expect women to leave to have families they wouldn't have to promote them for long* (106, male). With promotion opportunities for graduates becoming more restricted as they progress up the career pyramid, it could be argued that it is useful for companies to have a supply of female scientists and engineers who, because of their family commitments, were prepared to accept reduced chances of promotion to senior levels.

However, both mangers and employees drew attention to positive reasons for selecting women candidates. It was frequently noted, as others have found (Newton 1988), that the few women with technical degrees tended to be *determined pioneers*, with excellent academic records. Consequently, their drive and initiative could be considerably more impressive than that of men. Women were seen as more careful, diligent, methodical and logical in their ways of working, and their communication and persuasion skills were deemed better than those of men. Some managers and employees believed that women 'balanced' a department and that the presence of women in the work force might lead to improved productivity. One recruitment manager suggested that

> ...if anything, employers tend to over-compete for women. We would all hire on sight any decent female engineer without question, primarily for the balance of the groups. They [men] seem to work a bit better on the face of it when they've got female contact. It's just the chemistry of individuals. They respond to women being in the group better. They're cheery, they keep the group going in difficult times. Men try harder in the company of women.

Positive discrimination

In a labour market where applications from qualified women scientists and engineers are scarce, employers who wish to recruit women might be tempted to be more lenient with female applicants at the recruitment stage. Positive discrimination has been, and continues to be, a contentious issue in British society. The management interviews suggest that it has been a hotly contested issue within the companies we studied. Most managers were opposed to positive discrimination, citing the illegality of such practices under the 1975 Sex Discrimination Act. The setting of targets, whereby companies recruit a certain number of women within a set period, was equated with positive discrimination and also met with disfavour. Targets were associated by some with a lowering of standards which would only undermine the cause of equal opportunities in the long run.

A small number of employees of both sexes thought that their companies practised positive discrimination in favour of women, although it was not always clear how such discrimination operated. In some cases it appeared to benefit women at the initial stage of being invited for an interview. One employee argued that because women were under-represented they were given a greater chance to prove themselves:

> They did say that they felt that there weren't enough women so they did generally tend to look at women's applications a little more closely and they were more prepared to give a woman a chance of an interview just to try and even the numbers up (005, female).

This view was also expressed by some of the managers we interviewed. One personnel manager suggested:

> I think probably the tendency is for us, subconsciously, to look more favourably on girls who apply because we are conscious all the time of the need to encourage girls to come into engineering. I

think it is largely subconscious because we try to pick the best candidate irrespective of sex. There is probably a slight tendency towards favouring girls.

In other cases, positive discrimination was seen to occur when women were given high profile work:

> When I was sponsored I think there was definitely positive discrimination some of the time. I think unofficially some engineers here like to say that they've got a female working on the job, and positively would give her a higher profile than is really necessary (008, female).

Some employees saw positive discrimination as a good thing, recognising the benefits to be gained by employing more women, or seeing it as an expression of the company's good intention to overcome past discrimination. It was not uncommon for employees to support a mild form of positive action resembling encouragement of women rather than discrimination against men:

> This particular person would like to say he had female engineers who were doing well... I don't mind that, if it helps me to get on. There are enough problems, people who discourage it, so you should positively encourage some of it but I don't think there should be too much positive encouragement. You've got to be equal for the job with everybody else at the interview, you shouldn't get it because you're the woman (008, female).

In these various ways, then, a few of the individuals we interviewed favoured informal measures to redress past inequalities between men and women. The large majority of managers and a significant proportion of employees, however, expressed a dislike of positive discrimination. One male employee who felt that positive discrimination took place in his company said that as a result women had been recruited *in some cases to the detriment of good quality*. Women employees often resented any suggestion of positive discrimination because it implied that they had been recruited not on merit but because of their sex:

> Sometimes I think that they are biased towards women. It tends to make me feel that I got a job because I was a woman and not because I was academically qualified, and that does annoy me (049, female).

Company policy and individual prejudice

Despite equal opportunities legislation and company policies promoting equal opportunities, individual men and women employees from a number of the companies we studied maintained that, all else being equal, they – or others in their companies – would prefer to employ men rather than women. This position was justified by references to women's traditional responsibility for the care of their families and the potential conflict between work and family commitments. Women and men were perceived to be unequal in the commitment that they could give to a job. For this reason, some employees argued that recruiters were right to favour male candidates.

Employees also described other reasons for prejudice against women. In some cases, women were seen as unsuitable for engineering work:

> Maybe for some jobs they would be less keen to take on women, perhaps the dirtier jobs they might think that a woman couldn't do it quite as well or wouldn't be quite so suitable as a man (072, male).

> I think there's still a lot of men in industry who don't think women could do the job... that kind of stigma still exists and the only way [the company] is going to get rid of it is by employing more women in these kinds of positions (130, male).

> With computers it's a lot easier to attract women, but the hardware engineering... it's a man's job and if a woman wants to do it she'll have a really tough time (104, male).

Sometimes, women were not wanted simply because they were women:

> I don't think it boils down to a question of technical competence... there is a perception that women will behave in a different way when under stress, when they're suddenly frustrated, and that might cause ripples that we just don't want to have... they see it as suddenly there's one odd ball in the area, do we want it? (103, male).

The preference for male employees went against the companies' equal opportunities policies. Employees from all of the case study companies reported that individual biases and prejudices militated against the recruitment and progression of women in their organisations. The biases of individuals were often contrasted with the companies' aims of equal treatment in the recruitment process:

> The company is open to both sexes, but you can't escape individual biases (089, female).
>
> The company is just after the best engineers it can get... I can't guarantee that's true in every department because it depends so much on who's running each group (048, female).
>
> I think it [the company] takes a very positive attitude to employing women, as company policy. Again, I know of men who don't want to employ women (007, female).

Equal opportunities policies were sometimes identified with personnel departments, whereas engineers and line managers, responsible for implementing equal opportunities policies, were seen as more conservative and less concerned to see women being employed:

> Probably personnel... would like to see more women but certainly in design and engineering where I am I don't think that they would go overboard on it (002, female).
>
> Engineering in general is a bit of a man's club mainly because the men that run it are so conservative (123, male).

Reports of bias and prejudice among personnel with recruitment responsibilities were made by employees in nearly all of the companies we studied. Our evidence suggests that well-formulated equal opportunities policies were being sabotaged, effectively if not deliberately, by individuals in positions of power. Older, male recruiters in particular were considered to have prejudices against women scientists and engineers:

> There are some older managers who have not yet accepted the idea that a woman is as good as a man (103, male).
>
> Older management is prejudiced against women (061, female).
>
> [wanting to recruit women] would depend on the particular managers for each section, because when application forms come in they go to the heads of each department and they decide, and from what I can see, an awful lot of them prefer men (002, female).

Many women employees were dissatisfied about the way in which individual prejudices continued to influence recruitment decisions. Their views suggested that companies needed to do more to counteract the impact of such prejudices:

> If there's a particular man who you think will have an attitude problem then he has the problem, not the department as a whole.

> I'm not sure how men should be educated but I think the company needs to present to everyone what their policy is going to be as far as women are concerned, as far as any equal opportunities thing should be concerned... the people who work in the company need to be aware of that and anyone who doesn't like that can leave (010, female).

Interviewing potential recruits

Final-year science and engineering graduates who apply for a job with one of the companies we studied usually are interviewed on two occasions. The first interview, which invariably takes place on campus as part of the annual milkround, is informal and allows the graduate and the company to make some preliminary judgements about each other. At this stage, companies are concerned to find out more about a student's degree and particular interests, with a view to locating the student within their organisation. Potential recruits are judged on their academic records and on other qualities such as communication, persuasiveness, initiative and drive, planning and organisational skills. On the basis of this assessment, a second, on-site interview may be recommended.

Many of the companies we studied deliberately chose younger employees, both men and women, often with similar degree disciplines to the students, to conduct the first round of interviews. Young men and women, managers argued, could provide the most up-to-date information about graduate opportunities in the company, and were assumed to be more in tune with the hopes and fears of new graduates at the point of choosing their careers. Women interviewers additionally could answer any questions relating to the employment of women.

Most companies also tried to ensure that women managers were included in the on-site interview panel, in part to ensure that women candidates were not unfairly discriminated against and in part to show that the company did employ women in senior positions. It was assumed that women managers would not themselves discriminate against other women in ways that a male line manager might, and would be vigilant against unfair selection practices. For some of the companies, finding senior women managers to sit on interview panels was difficult, for the reason that few women occupied senior positions within their organisations. Therefore, many women candidates would still face an all-male interview panel. In the larger companies of the study, who had witnessed a gradual increase in the number of women graduates entering the company, this problem arose infrequently although it still had to be addressed. British Gas, for example, sent successful candidates from the milkround interviews to selection workshops at assessment centres and care was taken to include women interviewers in the assessment process. Their recruitment manager explained:

> We have worked as well at trying to get women assessors because that is another problem. It is all very well to have people at the milkround stage and then when you get to the assessment centre, if you are not careful, you have got a dozen grey-haired, 50 year-old [male] engineers who are making the assessment. Now that says a lot. So we have deliberately, over the last couple of years, tried to change that and we need someone with enough responsibility in the recruitment decision... So we have worked quite hard at getting assessors who are women.

The interviews we carried out with management revealed that in all of the companies under review steps had been taken towards formalising interviewing procedures in order to eliminate discrimination and to ensure equal opportunities as between men and women. Some companies were further along this path than others, but all were engaged in the process.

Nonetheless, the picture presented by employees suggested that, in practice, the types of recruitment process that were in operation potentially could lead to discrimination against women and other minority groups. Individuals retained the power to make recruitment decisions, and only a minority received training in conducting recruitment interviews. Very few had received training which dealt with the issues of equal opportunities.

Employees and managers in all of the companies we studied were aware of the tendency for women to be discriminated against in job interviews. Research has shown that questions about commitment to careers, or about marriage plans, have often denied women equal claim to a job, as have informal and indirect means of assessing candidates (Collinson 1988; Curran 1988). In light of this, the Equal Opportunities Commission has recommended that companies train employees involved in the interviewing of candidates on the provisions of the 1975 and 1986 Sex Discrimination Acts (EOC Code of Practice, 1985).

A small number of the companies under review provided extensive training on interview techniques including equal opportunities. Generally these were the larger companies of the study such as Esso, ICI, British Gas and British Aerospace. Interviewers from these companies were sent on courses in interviewing techniques; they also considered the ways in which people were evaluated and assessed as part of the interview process to ensure equal opportunities in practice as well as in policy. Esso, in liaison with the Pepperel Unit of the Industrial Society, had evaluated its interviewing procedures as part of a series of seminars on equal opportunities for company employees in 1986, with particular attention being paid to stereotypes and attitudes which may act against women. The company's recruitment manager explained:

> In all our recruiting and interviewing, we brief everyone on this whole subject of what's acceptable, what's not acceptable. There's still people out in this world who think it's acceptable to ask a women what her boyfriend does, if she's going to be free. What happens if your husband gets a job or your boyfriend gets a job? Will you really come and work for us? What's going to happen when you have babies and all of that? I hope that no-one does that here. If they do, I'd like to know about it. There's a conscious awareness that you're not allowed to ask questions like that anymore. You shouldn't. It's offensive and its not the name of the game.

From the perspective of the employees we interviewed, only Esso appeared to provide thorough training in equal opportunities issues for interviewers. Employees from this company could describe clearly the way in which such training was carried out:

> There is very clear guidance through practice and case-study type work – looking at when judgements are being based on the wrong data.

> They talk through the statistics on how many women, disabled and ethnic minorities the company recruits. We're briefed on any possible points that might stick out – things we're not allowed to ask in the interviews, things we're not allowed to say, biases we're not allowed to show.

> We have a maxim that really says 'you don't ask a question that you wouldn't ask a typical sort of British white male'... we encourage people to make sure that's what they do and we keep statistics of people we've interviewed.

> We have a three-day interviewing course which you have to go on before you're let loose on any interviewees.

Employees from Marconi Defence, ICI, STC, British Aerospace, British Gas and Plessey said that they had received training in interview techniques. This did not always mean that all interviewers in the company had been trained or that training courses were compulsory for people carrying out interviews. In some of these companies, employees reported that they had been given the opportunity to go on an interview course but had not taken it up, or that the training had not been provided on time. One employee said that he had requested training but had been reassured that he was doing all right.

British Aerospace had made use of external expertise in overcoming interviewer's stereotypes and employed an outside agency to train all of its interviewers on all aspects of the company's policies and practices, including its equal opportunity policy. Thus, one of its training managers felt confident in saying:

> If they have got in front of them an able engineer, it does not matter whether it is a man or a woman. That engineer will be processed right though and given every opportunity to join the company. We have trained them all, paid for them to have professional training, not only in interviewing but what the company is all about.

In a few companies, equal opportunities policy was not discussed, and recruitment managers felt that such a discussion would only generate controversy. Men and women *would be* interviewed and evaluated fairly; therefore there was no need to discuss equal opportunities. These companies, it could be argued, were denying that sex discrimination existed. The way in which they ensured that interviewees were fairly judged, irrespective of gender, was unclear, and appeared to depend upon the skills of individual interviewers.

There were two companies in particular where, despite company policy, hardly any of the interviewers we spoke with had received training. In one of these companies, women employees recounted their recent personal experiences as potential recruits when they had been asked a variety of questions about their future marriage plans:

> [the interviewer] was asking all sorts of questions like 'What happens when you leave to have children?' and you know, 'all the women we've ever employed have left'.

> [there was] cross-questioning about them marrying and having children, coming back to work, but not in a positive way.
>
> When I was at one of my interviews I was very careful to say that I wasn't going to have a family because I felt that that would stop me getting a job. There are some people around the factory who do give me that opinion that you are not worth them wasting their money on to train if you're going to have a family.

Moreover, in most of the companies, general interview training and guidance with specific reference to equal opportunities policies was provided in a less systematic manner to on-site interviewers than was the case with for those who carried out milkround interviews. The larger companies of the study were the most thorough about the provision of training for senior employees. But even these companies appeared to be less systematic about the guidance senior employees might need for interviewing men and women fairly. In some instances, it was left to the discretion of line managers to apply for a place on a course on interviewing techniques. Companies could not therefore guarantee that all senior employees engaged in interviewing had received interview guidance training, and could be even less confident about specific training about the company's equal opportunities policies.

Some of the companies had, then, formalised their recruitment procedures and had established clear criteria of the skills and competencies which they required from each candidate. This increased the likelihood that candidates would be judged as suitable or not on the basis of their qualifications and potential ability, irrespective of their sex or any informal assumptions and stereotypes held by individual interviewers. The variation between companies in the amount and quality of training given to interviewers suggests, however, that there remains a good deal that some companies could learn from the good practice of others.

Monitoring

All of the companies we studied monitored their intake of men and women graduates, as part of the central administrative process governing graduate recruitment. Monitoring was a new procedure for most of the companies concerned. The majority were still establishing their data sets, while a few others had been monitoring for durations up to ten years. Some companies merely kept final tallies of the number of men and women graduates they had employed in any one year. Others undertook more detailed monitoring procedures. The male and female ratio of applications received, success at first and second interviews and the final offer and take-up of employment were among the more detailed procedures followed.

The question of monitoring, like positive discrimination and the setting of targets, has been a controversial issue among those concerned with equal opportunities. Debates which took place among interested bodies outside the companies also took place inside. The recruitment manager at British Gas put the case for monitoring:

> I am one of those who argued that until you monitor, you do not know what you have got and you do not know whether you are discriminating. You then can hide under a cloud of 'we do not discriminate' but I believe in monitoring. I think that debate must go on nationally.

The statistical data gathered through monitoring procedures was analysed and made use of by some companies more than others. Companies who were active in this area included Esso, ICI, Marconi (Stanmore), Plessey, and STC. These companies were well informed about trends in male and female employment within their organisations. All of them were anxious to see that their recruitment of women matched national figures on the number of women graduating in science or engineering. If the figures did correspond, the information suggested that discrimination was not rife within the organisation. The recruitment manager at Marconi Defence Systems explained that monitoring was undertaken

> ...for our own sake, just to see that we are letting women come though and join us and the ratios are not too out of synch with what people are actually studying on the course. It would not appear that they are.

Other companies, however, suggested that little use was made of the data they collected. As one recruitment officer stated:

> I would if I had the time but we have so many vacancies here at the moment that we are falling over ourselves just to get them through the door. In terms of priorities, I am afraid that is not an immediate one.

It was generally held that the intake of women graduates in any one year could only mirror the number of women graduating with technical degrees from higher education. Managers, therefore, often saw themselves as passive agents and offered traditional accounts of

Women into engineering and science

labour supply which focused on the characteristics of individuals within the pool from which they drew (Rubery 1988:251). At the level of graduate recruitment, there was little they could do, they argued, to alter trends. Echoing the sentiments of other managers, the recruitment manager at ICI concluded that *however hard you try to select women, if they are not applying, you cannot select them.*

Part 2
Retention policy and practice

Current and expected skill shortages and projected changes in labour supply have brought staff retention to the fore as an issue which companies must address, or risk loss of competitiveness. Continuing skill shortages of course imply that managers will be likely to face increasing difficulties recruiting, retaining and replacing qualified, trained employees. Additional pressures on retention are likely to flow from the introduction of a single European market in 1992, as the mobility of scientists and engineers between countries is increased; and from the projected growth of the financial sector (Rajan and Fryatt 1988), which offers highly competitive salaries to technically qualified graduates.

Staff turnover is not always nor inherently problematic. A reasonable level of turnover may be desirable as new employees introduce new ideas and practices. Nevertheless, excessive turnover can present problems. Managers in the companies we studied referred to the economic costs of high rates of turnover. Companies stand to loose their return on investment when qualified, trained employees leave, and they incur further costs through the need to recruit and train new employees. Finding ways of retaining staff, and particularly employees with key skills, therefore makes good economic sense.

Managers in the companies we studied were reluctant to say that they had experienced any particular problems with the retention of scientists and engineers. Most graduate engineers consider a change of company after their first three or four years in employment. Often acting upon advice from their university lecturers, young employees invariably change jobs to order to gain experience of different types of engineering projects. This experience subsequently enhances their promotion prospects. Once their training is completed and they have acquired job-related expertise and experience, young engineers have the necessary technical skills to work competently, efficiently and independently. They are then an attractive proposition to other companies, and as managers suggested, exploit their position fully.

In the past, the loss of engineering graduates at this point in their careers had been accepted as unavoidable. The engineering industry has always experienced high levels of turnover, largely as a result of the favourable labour market position enjoyed by scientists and engineers. Managers were increasingly worried, however, about the the loss of engineers to jobs outside the industry. Engineers not only were becoming increasingly mobile between engineering companies, but also were entering a variety of general managerial positions in non-engineering companies, particularly in the financial sector. Managers in the predominately engineering, companies of the study reported that turnover was slightly higher among engineers than they would have wished. Turnover rates had been between 10 and 15 per cent per annum for the past several years; but the trend now appeared to be increasing towards 20 per cent.

Our study has identified three key areas that must be addressed by companies if they are to retain their valued young engineers and scientists. We examine these key areas in the following three chapters. First, opportunities need to exist for young employees to develop their careers and expertise in the directions they want through the provision of appropriate training (Chapter 4). Secondly, promotional opportunities need to exist in technical as well as managerial careers, and to be based on open and objective criteria that are recognised and understood by employees (Chapter 5). And thirdly, steps need to be taken to facilitate the continuing employment of women over their childbearing years and to ensure that women, including women with children, have equal opportunities for career progression (Chapter 6).

4 Training

Technical graduates enjoy an advantageous market situation. Their terms and conditions of employment in the first two years of employment at least, are highly favourable and include above average starting salaries and an emphasis on early training and professional development. In this chapter we assess the provision of training in the companies we studied, paying particular attention to the experiences of the employees we interviewed.

In line with company policy, managers were anxious to secure equality of opportunity in training for men and women alike. Given problems of labour supply, close attention was paid to the progress of all technically qualified graduates though the organisational structure of the companies, but the career progression of women scientists and engineers was subject to particularly close scrutiny. Our evidence suggests that only minor differences in access to training existed between the men and women employees of our study.

The provision of training

All of the companies provided technical and managerial training, and most provided training in the development of personal skills, such as interviewing, oral presentations, team-working and assertiveness. Methods of training delivery included internal and external courses, seminars and conferences, and on-the-job training. Many employees said that their choice of company had been influenced by the training opportunities which they perceived to be available.

The provision of training had changed in recent years in all but one of companies we studied. With one exception, companies no longer provided highly structured training programmes through which trainees spent a short period working in a variety of departments in order to gain experience of different types of work. Instead, after a short induction period, graduates were placed directly in jobs and, over the first two years of employment, were expected to receive both the general and technical training required to do that job. Only Westland continued to provide a formal training programme, seconding graduate engineers to a number of different departments in order to provide experience in design, engineering and manufacturing practices.

Managers gave a number of reasons for this change. The recruitment manager at STC suggested that highly structured training programmes were no longer appropriate because the company increasingly employed technical graduates from a wide range of disciplines with different abilities:

> Four or five years ago, we had a standard training programme but what we are finding from recruiting from the 90-odd universities and polytechnics around the country is that standards are quite different. Specialities are quite different and so are standard courses really. Some people are actually having to learn subjects within a few days and there's some people who have actually studied the subject in great detail... so there's a great disparity between the levels of knowledge of the candidates.

A training manager at the Plessey site at Poole discussed the preferences of graduate engineers and scientists:

> Many graduates, having had sixteen years of full-time education actually want a bit of responsibility and the chance to flex their muscles.

Many employees reflected his opinion in the reasons they gave for choosing particular companies.

Indeed, many of companies stressed in their company literature that new graduates would enter genuine jobs. Thus, while Ove Arup, for example, placed considerable importance on training in its graduate brochure, they also stressed that new graduates would enjoy early responsibility and manage small projects of their own. In place of the same training programme for all recruits, emphasis was placed on the need to provide individually-tailored training plans, agreed by the company and the individuals concerned.

This is not to suggest that the majority of the companies we studied did not have training programmes for their new technical graduates. Managers in all of the companies reported that each year's intake of graduates were placed on a short induction course in order to introduce them to the structure and organisational make-up of the company they had joined. Graduates were informed also of their terms and conditions of employment, the training which they could expect during the first two years of employment in particular, and the prospects for promotion in the long-term. Induction courses were seen as a way of orienting the individual towards the company, providing them with their first insights into the organisation. Employees in general valued the induction courses and expressed dissatisfaction when they missed out on this initial training. However, one or two commented that the induction programmes in their companies had been badly organised and a waste of time.

Once in their new jobs, graduates attended both general training courses and technical courses. All of the companies provided a similar range of general off-the-job courses which all graduates could attend. Such courses included report-writing skills, presentation skills and managerial training. Although some employees had received extensive managerial training, of high quality, many said that there was insufficient access to managerial training. One employee, for example said that it was *almost non-existent* in her department and that this hampered career progression. Other employees, also said that they had had insufficient training for a management role:

> The amount of training that we get is probably not the level that it should be for the amount of responsibility that's taken on (126, male).

British Gas had begun to tackle the problem of management training by setting up a standard training programme, compulsory for all new appointees to first management level posts, and leading to a diploma qualification recognised by a British polytechnic. Employees from British Gas were very positive about the new programme, and those who did not automatically qualify were queuing to be nominated:

> There's so many people want to get onto it and there's so many people been nominated by their managers that I'm probably a little way down the list.

The majority of employees said that they had received some formal technical training. Some had been sponsored by their companies to undertake extended programmes of technical training, such as external MSc courses. Others had attended short courses of job-specific training in technical areas. Some employees said that attendance at technical conferences and seminars was regarded as a useful form of training.

Most employees, however, had received a large part of their technical training from experience on the job. Good on-the-job training has the benefit of providing a supportive environment, where the employee is learning constantly and has opportunities to put theory into practice. However, on-the-job training may not be defined formally as training. Experiences described by one employee as on-the-job training were seen by others as *picking it up as you go along*. The following employee, for example, considered that he did not need to have training:

> For the sort of work I do, I don't think I need technical training... it's the sort of thing you have to learn as you go along, go to other people's labs if you need to learn a particular technique (083, male).

> They may not realise it but they're actually being trained every day in their working environment... when you are put into very new environments then you may think you're walking a tightrope but there's actually a safety net very close underneath you (093, male).

Several employees indicated that they would welcome formal technical training to fill gaps in the knowledge they had gained on the job:

> There should be a lot more technical training but it's very difficult to justify... most of it usually happens on the job whereas it would be better for people to get a more overall knowledge... so they could understand other people's problems (036, female).

Managers stressed that new graduates were encouraged to obtain professional qualifications. A number of training programmes were specifically designed to meet the requirements for chartered status as set out by the different engineering institutions. Plessey's training scheme, for example, was approved by the Institute of Electrical Engineers. The recruitment manager at Plessey suggested:

> We look at what their chartered requirements are and what their specific technical requirements are, so there is an individually tailored training plan for each individual.

Ove Arup provided new graduates with the necessary design and site experience in their first three years of work to meet the requirements for chartered status laid down by the Institute of

Construction Engineers. The three year training contract which Ove Arup agreed with graduate entrants was widely praised by employees:

> Arups agree to train you and you agree to train with them for three years – because they are a big firm they can afford to have in-house courses, so you effectively get all the training you need for the first exam.

> They pay for us to go on courses. We have to attend so many courses and collect what they call Chilver days... they give us the information so that we know what courses are available... moving us around so that we've got the experience we need to qualify for chartered engineer.

The programme appeared to benefit the company by providing an incentive for recruits and by reducing labour turnover, at least within the training period.

The allocation of training

Managers and employees reported that participation in training courses was open to all employees, irrespective of sex. All new graduates would attend the general courses on the companies' business activities and other similar induction courses. They would not necessarily all attend courses designed to increase technical expertise, as the need for such expertise varied across departments. But because training courses were equally open to men and women, managers felt confident in stating that their graduates enjoyed equal opportunities for training, and argued that there should be no reason why women would receive less training than men.

In the majority of the companies we studied, responsibility for endorsing the provision of training rested with line managers. As with recruitment, individual responsibility could result in individual biases determining policy and in individual competencies determining its effectiveness. The problems were succinctly described by an employee from British Gas:

> [It doesn't work] unless the manager himself had a clear view as to the sort of training individuals needed and programmed it himself. Without that, it inevitably resulted in people picking a course and going on it without a proper thought as to which course ought to be selected to follow which course. It often meant that people who perhaps were in desperate need of some aspect of training were able to miss out or opt not to do it... training is not merely something for the benefit of the individual, it can often be

for the benefit of the company, and just because a person doesn't see a personal need for it or in fact doesn't want to go, that really shouldn't enter into it.

Some line managers clearly were committed to discovering good training opportunities. However, the availability of such opportunities was sometimes badly publicised, and there were variations in the capacity of companies to supply the training which employees requested:

> There is no published information as to what is available, and as a result, a lot of people don't go on any. They don't know about it and therefore they can't ask for it (089, male).

> I have to rely on my section leader coming up with a list of courses and saying which ones are you interested in, but even if I say I'm interested in going I haven't always been sent on them (003, female).

The informal organisation of training allocation in most companies meant that a great deal was left to the initiative of individual employees. Employees from the same companies, of similar ages, presented very different pictures of the amount of training they had received. Those who shouted loudest seemed to be the most content with their training:

> I feel that the company has been very good for me personally. I am quite vocal in asking for what I need, and I do notice that not everybody gets what they want (056, female).

> I've made sure I organise my training. I look at training courses and go back to my supervisor and say 'I think I would benefit from going to that conference' or to that course and it's no problem (062, female).

> One of my abilities is to get up and get what I want rather than expect for it to come to me on a plate. And if you're prepared to do that within your job there are ample opportunities (117, male).

Women appeared to be well represented among the individuals who felt they had done well out of the informal system. However, while companies should not be criticised for encouraging and rewarding individual initiative and responsibility, it is not clear whether giving individuals responsibility for their own training will always be an effective method for creating a well-trained and well-motivated staff. At the least, individuals need to know where to look for training opportunities and how to apply for them. Ou evidence suggests that even when employees do take this initiative, their efforts may be blocked by unsympathetic line managers who place less value on formal training:

> I think my old department suffered from the old problem which they adopt in British industry, which is that they don't train up someone because they are frightened that they are going to lose them – that they will move on to another department or to another company. And, as I say, I certainly didn't have any training and I was very surprised that they actually let me go on a course (063, female).

> If you're a woman you have to really push yourself... I say 'have you any courses?'... I went and complained and the guy who had the thing sorted out said it was the [fault of the] engineering manager. So I went to the engineering manager who said 'it's not me, it's the other guy', so obviously someone wasn't telling the truth (061, female).

> I mean, [the company] do these training courses but it's a slight management problem, I think, in my department that causes the difficulties... my immediate boss is, I don't know, he seems a little reluctant sometimes... it's just his attitude (014, female).

Satisfaction with training

Employees from half the companies in the study expressed dissatisfaction with the amount of training they had received. In some cases, however, employees from the same companies were satisfied, giving some indication of the unevenness of training provision.

Training is costly to provide and financial constraints were sometimes responsible for limiting its availability. Employees from three companies said that they had been prevented from going on training courses because there was *no money*. In one company, it was reported that technical courses were given a priority at the expense of other types of course because resources were scarce:

> I feel one of the problems here at the moment is that there is not a very large training budget, so training courses that people want to go on are looked at very carefully, and the cost especially... the sort of training courses that I might want to go on now, not so technical but more to do with personal skills, don't tend to be approved so readily (053, female).

Budgetary restraints sometimes were responsible for restricting the availability of managerial training:

> I would have liked to have gone on a company management programme but it hasn't been possible to do that due to financial restraints (114, male).

Pressure of work was also given as a reason for not being able to attend training courses. This was the case with employees from three different companies and might, of course, reflect their own individual workloads rather than the general situation within their companies.

In general, employees seemed to prefer courses that were directly relevant to their work. An employee from Esso praised her training for being *specific to where your career is heading,* and for concentrating on technical skills in the early years and supervisory and management skills later on. Where there were criticisms of the quality of courses, these tended to relate to general training, such as that provided on the standard induction programmes, because it was not directly relevant to the employee's work.

The timing of training was important. Positive comments were made when training was felt to be well organised and well-spaced with regard to practical experience. Poor timing diminished the usefulness of courses for employees:

> When they put you up for courses they're often some way ahead and by the time you get to be on the course it's no longer relevant to what you're currently doing (123, male).

> A lot of it has been at the wrong time. I've been on several courses where I went because there was a spare place, which gave me some extra training but it wasn't particularly relevant... then two years later when you really need the skills brushing up on you can't go on the course because you've already been on it (006, female).

Poor timing was sometimes related to problems with training budgets:

> Training had been identified as a requirement of mine but as there wasn't a budget I couldn't go on it, and by the time I could have gone on the training I didn't need it because I'd learnt it anyway but very inefficiently (122, male).

> It's difficult to get on the right course at the right time because of budgetary problems (111, male).

Equal opportunities for training

Most managers and employees thought that men and women had equal opportunities for training. A few employees, however, drew ou attention to areas of possible discrimination. A male employee reported that men and women in his company did not have equal opportunities for training because there was a tendency to segregate them into different areas of work:

> The women don't get the opportunities in manufacture that they should, and probably the men don't get the opportunities in the business side that they should (130, male).

Access to training was felt to be more difficult for employees with part-time contracts. Some part-time employees felt that they were not entitled to ask for training, one remarking that:

> You can't ask to go on too many training courses because they don't have much of your time anyway (054, female).

Since it is mainly women who work part-time, differential access to training for part-timers might constitute a form of indirect discrimination. It would be good practice for employers to devise formal policies ensuring that part-time employees have fair access to training, on a pro rata basis if appropriate, and that they are made aware of their training entitlements.

5 Career progression

Companies need to provide good opportunities for career progression in order to retain qualified scientists and engineers. Our interviews with employees indicated that many of them were prepared to consider leaving their companies if they felt that promotion was too slow or that promotional opportunities were limited. There was evidence that graduates set themselves targets for promotion and that, if these were not met, some at least would have little hesitation about uprooting themselves and pursuing their careers elsewhere:

> If I don't get to a certain level in about three years or more I'll change companies. I'm not going to sit around here and work for years without getting anywhere (068, female).

Managers acknowledged that promotion was a key issue in the retention of scientists and engineers. They recognised that after three or four years engineers are eager for promotion and may become frustrated by their lack of mobility through the company. Our study provides evidence that companies were taking steps to facilitate the career progression of scientists and engineers more explicitly and directly than may have been the case in the past.

Opportunities for promotion

Management start to guide graduates though the hierarchical structure of a company almost as soon as they enter the organisation. As new recruits, considerable time and money is devoted to their progression to managerial positions in particular. The larger companies of the study were likely to have a team of senior managers who devise long-term career plans for their graduates, identifying jobs for them at an early stage in their careers and charting their paths through the company to reach their final position. The predominately engineering companies were increasingly engaged in similar practices, and had begun to identify career paths for their graduates.

The early careers of scientists and engineers within the companies we studied varied considerably. Esso, for example, moved its new graduate employees to different jobs within the organisation every eighteen months to two years. An engineer who had started work at the oil refinery at Fawley could be working in a contracts or supply department in head office within a very short time. This situation was found also, although to a lesser extent, in other larger companies where they wished to develop *managers* rather than technical specialists.

In the predominately engineering companies it was also possible to describe a typical pattern of graduate career development for men and women. Over a two-year training period, new graduates would develop both their general and their technical skills on a variety of different projects, increasing their level of competence and job performance. Enhanced competence would be rewarded with an almost automatic progression up the occupational hierarchy. As part of this process of upward mobility, a graduate would steadily take on more responsibility. By their mid- to late-twenties, a typical engineer would be directing all or part of a project, increasingly supervising the work of others and spending an increasing amount of time on managerial tasks. The time spent on detailed technical work would accordingly be reduced.

However, whatever the size of the company, in all but one of the organisations we studied, promotion opportunities were relatively abundant in the early years of a graduate's career. A Boots employee described the system for recruits to her company:

> They tend to take people on at 'one' and once you gain a bit of experience in the company it's generally accepted that you go up to 'two'... The 'one' is a sort of trial stage.

Early promotions tended to be withheld only if performance was unsatisfactory:

> Most graduate engineers are made senior engineers within three or three and a half or two years. It's not so much a reflection on how good you are, it's a reflection on how poor you are if you don't make it (059, female).

Opportunities for promotion for scientists and engineers, as would be expected, gradually diminish as they progress upwards in their companies. Senior managers, while judging employees on their performance at work and potential for the future, must therefore be increasingly selective as they guide peoples' career paths through the organisation. They must begin to distinguish, for example, between those graduates who will occupy middle managerial positions and those who will occupy top positions. In many companies, the number of senior positions was more or less fixed, and employees had to compete for the vacancies that arose:

> Promotions are always limited to what position has become available and positions become available due to other people's promotions or due to people leaving the company (084, male).

Limited promotion opportunities were more likely to occur in smaller companies with fewer senior positions to fill, in companies where there was little mobility between functions and departments and, more generally, as one moved up the scale of seniority. One employee related the problem to the state of the wider labour market:

> The employment market in general was tight for a long time during our period of stagnation, and recently the last three or four years it's certainly got a lot more buoyant, with a lot more job opportunities, and that's led to people moving out (012, male).

Sometimes the lack of vacancies meant that employees felt that their achievements were not being recognised:

> Unless they restructure this particular division I can't see promotion coming for a long time, purely because the people above me aren't that much older than me and they're not going to move (080, male).

> There doesn't seem at the moment any way you can be promoted on your own merits, you really have to wait for the position to be vacated and then they choose the best person for the job (015, female).

In a few companies, the career structure was so blocked by lack of mobility that employees felt they had to leave, or threaten to leave, in order to get promotion. Some employees viewed such practices with cynicism, but there also was evidence that they had given serious consideration to these strategies:

> A good way of getting promotion is to get another job for six months and then come back, you can usually jump about two grades (067, female).

> Our boss is fairly bad about getting people upgradings unless you threaten to change departments or something... [then] he actually

> realises he needs you to stay on and that's the way you get upgraded (069, female).

Although some employees who were considering leaving their companies did hope to return after a period of time to a higher grade, others did not intend to come back. The most frequently cited reason for leaving was dissatisfaction with the opportunities for career development.

Opportunities in technical and managerial functions

There has been widespread recognition of the lack of opportunities for scientists and engineers to progress on the basis of their technical skills. Engineers with scarce skills have had to leave these skills behind in the pursuit of career advancement. Accordingly, the Finniston Report (1980) recommended that companies should create technical career ladders equivalent to managerial ones. Managers acknowledged that career progression tended to be through management positions, and suggested that this situation would have to change if engineers who enjoyed using their technical expertise were to be retained:

> The route to the top is the management route. I think that has hitherto been the case and it is recognised that rewards have come with going through the management scheme – benefits and salary, car and high status. A lot of our very good technical people have eventually moved across to the management ladder. We are addressing the whole question of the career development of engineers. For those who want to stay on the technical side, their careers must be equally assured and attract all the same benefits as if they were moving up the career ladder. We are tackling that issue. We have got to. We rely so much on those technical people for what we do.

With one exception, all the companies in our study provided some opportunities for promotion via a technical career path. In most companies, however, the managerial route was seen by employees as offering greater opportunities. Technical ladders tended to be narrower than managerial ones, offering promotion to only a small number of specialist positions:

> I was the first to be promoted [on the technical ladder] for eight years whereas typically managers are being promoted – there would be several every year... the next rung of the technical ladder is not impossible, but exceedingly rare (098, male).

Technical career ladders also were shorter than managerial ones. The highest positions in the companies were occupied by people with managerial responsibility:

> The more senior you get, the more you have to influence work, therefore to influence work you must have management skills of some description (081, male).

In several companies there were complaints that recognition and rewards were inferior on the technical side:

> People on the supervisory management track are valued by the company far more than the technical people... technical experts are valued by their own managers but the reward structure suggests that the company doesn't value them equally (092, male).

The lack of rewards for technical specialists discouraged employees from pursuing such careers and made it difficult for companies to retain valuable expertise:

> There's one area of the promotion/career development side of the way we operate that I don't think is good, and that's the way we reward the foundation people, the people who are very good at their scientific and technical jobs but who don't show management capabilities. We do have a tendency not to do very much for them, we don't really have a very strong parallel ladder or promotion ladder... which I think is a problem that we will have to address if we're going to keep good quality people actually doing scientific or engineering work (079, male).

Employees who had chosen engineering as a career because they liked technical work, expressed regret at having to leave this type of work behind in order to gain promotion:

> Most engineers, as they go up into management, get disappointed that they are not able to use the things they enjoy doing, the things that got them into engineering in the first place (105, male).

Some of the employees we spoke to did want to pursue careers in management; others, including substantial numbers of women, would, all things being equal, have preferred to continue with technical work. The ideal situation, for many employees, was a combination of technical work and managerial responsibility.

Assessment and career progression

Most of the companies used formal systems of assessment to grade their scientists and engineers. These frequently involved the setting of yearly objectives for each employee. An appraisal session would focus on whether or not employees had achieved their objectives, the reasons for non-achievement, and positive and negative qualities of employees. Training needs would be identified and the employee's performance would be assessed in relation to that of his or her colleagues.

Line managers played a crucial role in performance evaluation of graduates. Managers in all of the companies we studied reported that graduates were judged according to clear objective criteria which related solely to job performance. Accordingly the extent to which previously-agreed aims and objectives had been met was the major criterion upon which graduates were evaluated, although questions of initiative and drive were also considered. A manager at STC explained:

> It's really driven very much by his or her abilities. We move people through the career structure at whatever speed the individual is capable of. The vast majority move at a reasonable pace while a small number of people who are very high fliers, really cracking every job that they are given and performing dramatically better than everybody else, get moved up the grade structure very quickly. There are a very few people who perhaps stagnate at the lower end.

Assessment and evaluation of performance by senior line managers and directors determined, to a very large extent, the progress of graduates through the organisational structure of the companies. At the same time, managers in all of the companies reported that they increasingly took graduates' own career plans and aspirations into account, jointly agreeing and devising training within this context. As part of the review procedure, which occurred as many as three times a year within some companies, new graduates were asked for their assessment of the training they had received and their future training needs. These topics would be discussed with regard to the individual's career aspirations and the ways in which the company could facilitate progress to reach his or her desired objectives. In other words, graduates were actively involved in the assessment of their own work. The recruitment officer at Esso explained:

> Every year, we have an appraisal programme in which the employee plays a very active role. The employee has to fill in a form saying what they have done over the past year, what they are going to do next year, what training they have had, what training

they want, what sort of future they foresee. It's very much the individual playing a very active role in his or her appraisal.

The participation of graduates in the appraisal system provided companies with information on their employees' career aspirations. A manager at one of the Plessey sites suggested:

> We do treat all our people as individuals. We have a formal appraisal for each individual once a year so we know quite clearly what their career aspirations are and their goals. As a result of these documented discussions, we try and shape their careers in a way they want it to go because they are a valuable and scarce resource. It's recognised that if you push them down the wrong path, a path they don't want to go down, they'll just go.

Managers in the larger companies of the study, like ICI, were particularly likely to report that their methods of assessment were based on clear and objective criteria and took into consideration only individual ability and merit. Managers wished to ensure that evaluation criteria were systematically applied across the whole company, in order to secure equal opportunities for men and women. Managers argued that biases introduced by informal methods of assessing performance, based on individual attitudes and opinions, could only be eliminated by a highly-structured, formal and systematic appraisal process. They wanted, moreover, to see more women involved in the process of assessment to guarantee further that discrimination would not occur.

Employees generally supported the use of appraisal and grading systems, which were seen as useful in helping them to assess their own performance and to plan their career progression. Their comments suggested, however, that the quality of appraisal schemes, and the effectiveness of their operation, varied considerably among companies. One employee described the appraisal system in her company as *virtually inoperative*:

> The appraisals actually don't matter very much, or haven't in the past. It's been a waste of time. There's a form that has to be gone through but all that happens is that the member of staff gets the appraisal on their personnel record and it disappears without trace (012, female).

Employees from two other companies felt that the systems for structuring career progression were misused. This defeated the object of having an objective system and caused resentment. In one case the problem was related to the lack of a well-defined career structure; in another, to the fact that managers did not adhere to the system which did exist:

> The so-called structure changes from month to month depending on who they want and who leaves (089, male).

> There is no set way for promotion within this company. People misuse the system. There is a system, there are set rules, but nobody abides by them (064, female).

Employees in three companies had had difficulty in gaining access to information about career progression. One senior employee complained that posts were not advertised and that individuals were given little help in planning their careers:

> Everything is done apparently in secret, behind closed doors. There is no presentation of what jobs there are available and so there is no advertising of middle management and senior management jobs. There is a statement that says that people are responsible for their own careers but unfortunately there's no help to let them manage their career (102, male).

However, in two of the three companies where employees made complaints about the availability of information on career progression, other employees had been able to describe the system in quite a detailed way.

Criteria for promotion: the employees' views

The criteria required for promotion was an issue that employees had considered when planning the development of their own careers and watching that of others. Where managers had emphasised objective criteria based upon ability and job performance, employees brought other influences into the discussion, such as length of experience, commitment, personality, and good luck. Often, the requirement was felt to be a combination of several criteria rather than just one or two. The following gives a brief account of the different types of criteria emphasised from the perspective of the employees we interviewed, and considers the effect which they may have on women's opportunities for career development.

Ability and performance

Ability and job performance were widely regarded as important criteria for promotion by managers and employees. Sometimes it was sufficient to demonstrate competence; in other companies,

65

performance had to be not just good, but better than that of others in the employee's peer group. Managerial potential was often of particular importance:

> The ability to do the job and show you've got the skills to develop products, be conscientious and know what you're doing... so the management have an overall impression of you being competent (085, male).

> To be promoted I think you have to be able to take problems on, to be able to work on your own, be independent and show ability to supervise other people, take decisions... they're looking for you to be a manager, to take responsibility (090, male).

The ability and confidence to take on responsibility, run projects, and make decisions were highly-valued qualities which often featured in performance appraisal schemes. Many employees recognised that they were absolutely necessary for progression:

> You really need to have increased the responsibility in your job. So you can be working very hard... but if you haven't actually got decision-making responsibility or responsibility for getting things done that are critical to the company then you don't get recommended for a grade increase (029, female).

But merit was not universally regarded as an important requirement for promotion. One employee stated that *very few people actually get moved up through merit* in his company, a view supported by at least one of his colleagues. In this traditional engineering company, experience and length of service were regarded as the criteria determining promotion opportunities for the majority of employees, although a few high fliers had been known to move more quickly through the ranks. This resulted in considerable dissatisfaction among younger employees who complained that the criteria applied generally slowed down their rate of progress and led to stagnation.

Personal qualities

In nearly all companies, employees believed that personal qualities were important criteria for promotion. Some of the desired qualities were those traditionally seen as masculine, for example, being outgoing, forceful, and prepared to speak out. Some of the companies sent male and female employees on assertiveness courses to help develop these skills. One male employee thought that lack of these qualities put women in his company at a disadvantage:

> Competition for promotion and advancement is quite strong... there's a lot of forceful pushy blokes around and the girls that we work with don't seem to have the same sort of pushiness to compete (115, male).

On the other hand, qualities traditionally characterised as feminine, such as cooperation and the ability to work in a team, were also seen by employees as desirable, as were communication skills, particularly those of persuasion and diplomacy:

> It's the people who can spot the time to say something... to make their point, and to do it in the right way (094, male).

Candidates for promotion had to persuade their managers to recognise their abilities, responsibilities and achievements. Frequently it was not enough to be doing the job; you also had to be *seen to be doing the job* (091, male). Visibility was an important promotion criterion mentioned by a large number of employees:

> If you want to get promoted faster or further then you have to not only do your job well from the point of view of technically completing projects but you have to probably take the initiative to impress the management with your abilities (112, male).

Women were sometimes claimed to have an advantage over men by virtue of their greater visibility as members of a minority group in engineering. Some women agreed that this had been the case in their careers:

> There aren't many women, it's much easier to appear good just because you're in the minority. People tend to remember you... if anything, my promotion chances might be slightly better because I get a lot more high profile things (021, female).

There were a few suggestions that male colleagues resented women's greater visibility, but some men recognised that there also were disadvantages:

> Because she's a female, because she's a minority, she stands out more, so people observe her work and take note... that's a danger in a sense that if she does a bad job it will get more noticed than if I did a bad job, perhaps (117, male).

The opposite point of view, put forward by women from several different companies, was that women were overlooked rather than singled out for notice, that they were not expected to achieve as highly as the men and that, consequently, they had to work harder to gain the recognition that was due to them:

> You have to be better at your job and you have to make yourself more known and be seen to be better at your job than a man (066, female).

Commitment to the company

Competition for senior posts was intense in all the case study companies. Many employees considered that a combination of the criteria discussed so far – ability, experience, personal skills, and job performance – would not be sufficient for promotion to senior positions. In addition, candidates would need to demonstrate the right attitude to work: a desire for promotion, dedication, and commitment to the company:

> Your attitude to work is monitored quite closely... it will show if you're keen... [if] you're dedicated to your work, and how much you were willing to take on (049, female).

One senior employee claimed that there had been occasions when employees of alleged lesser ability had been promoted because they had been determined to succeed:

> They had this big strong desire to be in the more senior position. I think that's very important to the company because they are fairly stressful positions, they're highly responsible, and unless someone has that desire than there's not really any point in moving them forward (092, male).

Desire for promotion, and dedication to the company, like performance and ability, have to be communicated and made visible to senior management. The most common measure of an employee's commitment to the company was the amount of time dedicated to company work. In many cases candidates for higher positions were expected to demonstrate that they were willing to give up their own time, on a voluntary basis, in order to progress their careers. A number of employees believed that higher promotions were very unlikely to be achieved by employees who restricted themselves to working their contracted hours, commenting that *you can't work a nine to five job* (028, female) and *higher jobs do demand higher commitment* (059, female). Another employee reported:

> It's quite competitive and therefore you've got to be seen to do the right thing at the right time, which means endless overtime, being at the right place, and a high profile (044, female).

Although senior jobs carry increased responsibilities and sometimes increased workloads, some employees implied that extra hours were worked simply in order to demonstrate commitment to the company, and that this was as important as getting the work done:

> I think as you rise higher the company has got to see that you're putting in a lot of time. In some ways it's a bit of a sham because it could just mean that you're inefficient. If you're having to do a lot of work but do it in your basic time then there might not be any promotions (121, male).

An employee was considered to be dedicated if he or she could show that *the company comes first*. The sacrifice of personal time was a way of demonstrating this:

> To go to the level beyond that I think it is absolutely essential that you are prepared to be a company man [sic] and perhaps sacrifice time at the weekends, certainly work very late in the evenings (098, male).

> I've been in a senior position for a long time now. I'm not entitled to overtime but the job requires significant numbers of hours. Typically I put in sixty, maybe more hours a week, which I don't mind doing but it is galling to go home at the end of the week and find that the family does suffer from it. They're getting very little by way of return, very little compensation if you like, for doing without (121, male).

Good luck

Inevitably, perhaps, luck was perceived to play a part in deciding who received promotion. One type of luck involved *being in the right place at the right time*:

> There was a job available that I would have liked to have done and I was prevented from being considered for that job because I could not be released from my present job (033, female).

> I tended to be in the right place at the right time. Things slotted together quite well. One project would finish when another one was about to pick up and I would be available and able to move on to it (109, male).

> If we get a project that comes into the department and it takes off and you fly with it you're going to go very far very fast. If

everything you work on stops after six months, then you are never going to get a chance to shine (005, female).

Luck could also involve *getting the right boss*. In the organisation of promotion, as in the organisation of recruitment and training, there was evidence that power held by individuals could be used to provide opportunities for certain employees at the expense of others:

> There are so many little groups... and depending on how your managers apply the rules you find the people working in that group suffer or benefit... it's a lot to do with personalities and personal views and it's not structured at all (089, male).

> If you've got a good manager who will progress your career then you'll get through (099, male).

So long as prejudice against women continues to exist among science and engineering line managers, the influential manager can present a threat to women's opportunities:

> If the senior managers want someone to have promotion and to get on they'll find a way, they will twist the rules and they don't often do it for women. And you see it happening every now and again for the men (020, female).

Equal opportunities for promotion

Few women occupied senior managerial positions in the companies we studied. At the time of our interviews, none of the companies had a woman executive on the board of directors. All of the managers stressed that the scarcity of women in their top jobs was a consequence of the only very recent entry of women into careers in technical fields. Managers argued that many young women employees were now progressing though the lower levels of management, and that they would occupy senior position in the near future. The success of company policy could only be judged in the long term.

Against this background, a number of the companies under review were tackling the absence of women from top jobs as part of their commitment to equal opportunities. The companies emphasised their desire to see women enter managerial positions and some managers described women as having particularly good management skills. Referring to a management course run by the company for engineers wishing to embark upon a managerial rather than a purely technical career path, a manager at STC noted:

> We've taken mixed male and female courses. Almost without exception, the females perform better than the males from the assessment centre. It is often known, before the course starts, that the females will, in fact, be better managers. They seem to have many traits that are better suited to managerial jobs. They communicate with a good sense of expression, if you like, they speak well; and they do get the group together more than males tend to do.

Several companies were reviewing their assessment and evaluation procedures to ensure that no criteria, such as age or length of service in a job, discriminated against women. Stress was laid on job performance and potential as the most important criteria for assessment and promotion. The introduction of comprehensive appraisal schemes for the assessment of performance and potential was felt to enhance women's chances for promotion.

But managers also noted that other factors do come into play for promotion to the top jobs in their companies. The most frequently cited criterion, across all ten companies, was the need for a broad range of work experiences within the company. Senior managers, it was argued, are expected to represent the company and in order to do so, they must have experience across a diversity of jobs.

Line managers and senior managers determine which graduates will receive the necessary scope to facilitate promotion. Given this, senior managers within Esso, for example, had reassessed their assumptions about the types of role and jobs that women can occupy in the senior levels of the business – the implication being that not all jobs had been open to women in the past. As part of the company's equal opportunities policy, however, Esso was committed to seeing that women were considered for all positions within the company – an important advance for women if they are to achieve the broad experience needed for promotions to senior levels. As a consequence of this policy, the company had promoted women to managerial positions in predominately male, blue-collar sections of the company.

Managers reported other examples of positive action. Westland, along with other companies under review, supported a variety of external courses of special interest to potential, and already-established, women managers. The company supported women graduates who attended these courses, and provided in-house personal assertiveness courses for potential women managers (just as they provided special confidence-building courses for young women employed through the Youth Training Scheme). The aim of these courses was to counteract the early socialisation of women as well as to develop their managerial skills. Similarly, British Gas and ICI were in the process of developing special training measures for women as a way of actively encouraging women to develop long-term career plans and to pursue the training and promotion opportunities open to them within the organisation.

A substantial number of employees felt that there were equal opportunities for promotion. Some said that they were not sure, or that there were too few women in the company to draw any conclusions. There were, however, employees in nearly all of the companies who argued that it was more difficult for women to be promoted. Various reasons were put forward for this, some cultural and some to do with individual biases. In some instances, employees offered similar explanations to those given by managers for the under-representation of women in senior posts.

One explanation upon which both managers and employees were agreed was that women had only recently begun to enter engineering careers. Most women engineers were comparatively young, without the experience necessary to have reached senior positions. Some employees suggested that the time was arriving when more women would have that experience and should thus be in line for promotion:

> Historically, it's only now that the generation of women are actually reaching the stage where they want to go on to senior management (030, female).

The movement of increased numbers of women into positions where they might aim for management means that it will be important to monitor the performance of women over the next few years, to assess whether they are gaining the promotions which might be expected. A number of the companies were paying close attention to the career progression of men and women graduates through monitoring. In doing so, senior managers were looking very carefully at the reasons why women might fall behind their male counterparts, why more men than women might be on the fast track for promotion and why a man might be promoted to a particular job if a women of equal ability was also available. A manager at ICI, for example, reported that they had moved towards monitoring the ratio of men and women appointed to vacancies which arise within the company, in order to track the progress of men and women through the selection process. It was believed that monitoring the promotion procedures and mobility of men and women within the organisation would lead to more equal opportunities for women. In addition, the statistical evidence of unequal opportunities provided by monitoring was seen to be a powerful tool which managers could use in their attempts to translate policy into practice.

Esso was the only company where employees talked about the results of equal opportunities monitoring. Esso had recently implemented an equal opportunities training programme, stimulated by the results of the monitoring, which had shown that women generally had a slower rate of promotion than men:

> They looked at women's performance and men's performance and then charted their progression. They took graduates from, say, 1980 and looked at it and the women should, in theory, have done as well as the men. In every case, the women were several salary groups behind the men. That was the first time a lot of the senior managers saw that it does exist and that it is something we have to be aware of.

A second explanation for the poor representation of women in senior position was put forward by several women employees who felt that they were disadvantaged by the male cultures of their companies, and of the industries within which these companies operated. Women did not fit naturally into such a culture, and it was felt that this could lead to discrimination.

> I feel that the company has a *grey men in grey suits* culture, it's not ready to change its culture. It doesn't want to change its culture and it exists in an industry that has a *grey men in grey suits* culture (056, female).

> You're looking at clones that will perpetuate people and nobody can see a woman in the chairman's position, therefore, subconsciously women aren't promoted as quickly (024, female).

Several employees expressed disquiet at the potential reactions of customers at being faced with a woman manager.

> I'm not sure how customers would see a lady running a division (037, female).

> When we get visitors in, putting the female manager in front of them doesn't exactly help at times (126, male).

Individual prejudices could also lead to discrimination against women. Although most of the companies had adopted, or were in the process of adopting, objective assessment schemes, our employee interviews suggested that there was still scope for individual preferences to be the deciding factor in promotion:

> I think our director at the moment likes having a lot of women around. But I don't think he likes, I don't think he likes promoting them (042, female).

> Promotion is a difficult one. I think it's still not – for a woman to be promoted into a manager's job would still be a bit of a struggle. You do need to have a senior manager who will back you and who will argue your corner for you... when I look at the higher levels in the company, the women aren't there... [it all goes on] behind closed doors (103, male).

> [why women do not get promoted] has obviously got to be because of the people who decide who is eligible for the next promotion. Whether consciously or subconsciously – and I think a lot of it is subconscious – [some managers] just don't seem to think of women as full members of the company (066, female).

A few employees felt that men would resist being put in the position of having a woman boss, and there were suggestions that a woman would not be able to command the same respect as a man:

> The thought of having a woman boss to some people is quite frightening (053, female).

> There may be some prejudice if she was to take up a management position... the managers above that may feel that this isn't a very good situation because she's not going to get the respect from the men working under her (086, male).

However, the most common explanation given by employees for the scarcity of women in senior posts was that many potentially eligible women left work to care for their families. Some had returned but were several years behind their colleagues; while others had not returned:

> I don't know very many women who've got very high up... Probably because most of them leave to have families and once they come back then they're that much further behind (003, female).

> I think there's a lot of drop out for ladies, they go off to have children and then they don't want to come back. That's their decision and I think that's probably why you don't see as many women higher up the company (099, male).

The level of commitment demanded by some companies of their employees put considerable pressure on the employees' personal and family lives. This was widely perceived to make promotion more difficult for employees, particularly women, with family commitments:

> It's difficult for men in their domestic arrangements and it puts pressure on their marriages. So for women, if there's very much male-female stereotyping in their family life it would make it very difficult indeed (028, female).

> Women who start families and continue with families don't appear to go through the system... a woman has to decide that she's going to be a career-based woman and not a family-based woman (102, male).

Women with children (or with other caring responsibilities) are put at a disadvantage by the requirement that candidates for promotion should give first priority to their careers. A number of women had been discouraged from seeking promotion because of the extra commitments that advancement would involve:

> I've got two children... if I really wanted to just work, I could stay late and I could be at a higher level doing greater things, but no, I don't want to do it (026, female).

Another reported that she wanted further promotions, but *not too high*:

> I wouldn't like to get so high up that I end up doing a sixty hour week and have to take work home (062, female).

And another described how she was planning to delay having a family in order to commit herself to her job:

> If I had several young children I would not really think that I would be suitable for a job that may involve long hours of overtime and being called out to look at problems and solve problems on the plant twenty-four hours a day... it is unfortunately the technical job that comes at the wrong time which is why I would be thinking about not starting a family until my early thirties (013, female).

Several employees pointed out that it was not the presence of children which made career progression more difficult for women but the commitment that they felt it necessary to give to their children:

> If you are able to have a family and also a job with quite a lot of responsibility then you will not be turned down because you are a woman (031, female).

> As long as you're doing the job it doesn't matter to the company if you're a married woman with children or a man with children (023, female).

Some women, who had organised ways of reducing their commitments at home, had managed to have successful careers and a family. But there was no evidence to suggest that women scientists and engineers are any more likely than other women to be married to men who were prepared to take on an equal, or greater, share of childcare responsibilities. Mothers usually take on the primary responsibility for childcare and they may need to arrange for time off when their children are ill, when normal childcare arrangements break down or in other emergencies. Although some employees said that their companies were *generally compassionate* when staff had childcare problems, the majority of employees who commented on this issue thought that women's careers were likely to suffer as a result of requiring occasional time off:

> They would enjoy the same opportunities providing they weren't requiring time off to pick children up... if they were making special demands then the circumstances would change (125, male).

Several women said that they had been forced on occasion to pretend that they had been ill, because their companies frowned on the idea of granting leave to care for dependants:

> Even if she came back, I think it would be held against her to a certain degree because the company isn't that flexible at the moment to allow her to perhaps have time off when the kids are ill... that is always going to go against the current grain of working (015, female).

Unmarried women and women without children were affected by the assumption that the family would be a woman's first priority. Some individuals saw all women as inherently unreliable employees because of their actual or potential family responsibilities:

> Obviously it is more difficult for the company to get a woman heavily involved with a project because if she turns round in six months and says I'm leaving to have a kid, obviously that's going to mess everything up (119, male).

> If there was a woman in the company then it would usually be up to her to take the kid to school and bring the kid back from school and that could affect her at work. It could certainly affect her at work if... she had to go away for a couple of days either to Europe

or [elsewhere]... I think a woman would have a hard time trying to do all the things you are supposed to do to get promoted (110, male).

Only a small number of individuals reported that they held such views themselves, but at least one employee felt that they were held quite widely:

> Any top manager is thinking, well, is she going to go in a couple of years' time and have a baby? (121, male).

As a result of these assumptions, some women felt that they were being judged, not on the basis of their actual performance, but on the basis of how motherhood was expected to affect their performance. They felt that they were being denied the chance of proving that they were capable of fulfilling the requirements of higher positions. One woman described her experience of this:

> I think they've actually plateaued me because I've had the baby and I feel a bit stifled by that. I think they make the assumption that you can no longer do such a demanding job and I think it's going to be a lot harder to actually be promoted now (030, female).

A number of employees thought that the best way to break down resistance to employing women in senior positions was to provide proof that individual prejudices were unfounded. Women who are promoted to senior positions are often burdened with the responsibility of proving themselves equal to men. Their visibility as managers is much greater than the visibility of women scientists and engineers. Our interviews suggested that the performance of senior women was observed closely, both by managers and by employees working under them. Their successes and failures were particularly unlikely to be ignored. Women who are successful in senior positions can set precedents for promotion policy, and are seen as opening the gate for other women. A male employee described the way in which a woman manager in his company appeared to be regarded as a test case:

> The woman who did my last job before me moved into a line management job on the site so we're now seeing all the old arguments proved wrong. But they do have to be proved wrong before you can get a better acceptance that what has traditionally been a male site with all the men doing the heavy work, that you can actually have a woman in charge co-ordinating people and be respected (103, male).

Note

1. There were a small number of men with families who were beginning to place limits on the commitment that they were prepared to give to their careers. Although these men were a minority of those we interviewed, their comments are worth noting. Some were resisting the demand to work large amounts of overtime:

    > Having children has refocused my targets. I'm not as ambitious as I was three years ago... that's because I recognise that I don't want a job which is going to keep me here until eight o'clock at night (092, male).

    Other men said that they would not accept jobs which meant moving house or travelling abroad for long periods of time, because of the disruption that these things would cause to their families.

6 Retaining women engineers and scientists

In this chapter we examine company policy and practice in relation to initiatives directed specifically at improving the retention of women scientists and engineers during family formation. In response to problems of labour supply and acting in accordance with their equal opportunities policies, many of the companies we studied had developed enhanced maternity leave provisions and had brought in career-break schemes for women who wished to leave work for a period in order to have families. Other companies were considering the feasibility of part-time working for women in senior and technical occupations. We assess the effectiveness of these and similar initiatives and address in particular their implications for the promotion prospects of women. Our evidence suggests that much remains to be done by companies in order to retain women's skills during family formation and to ensure that opportunities for career progression for women are maintained.

Over the 1960s and 1970s, women increasingly took shorter spells out of the labour market between births and returned to work more quickly after completing their families (Martin, 1986). There also has been a marked increase in the extent to which women return to work within one year following the birth of a baby. In 1979, about one quarter of women in work during pregnancy were back in work within eight or nine months after the birth; by 1988, the proportion back in work within nine months after giving birth had all but doubled, while the proportion back in work full-time had trebled (Daniel, 1980; McRae, 1991).

These changes in women's behaviour during family formation have implications for both women and employers. Childbirth has commonly constituted a major disruption to women's working careers. If women are able to return, not simply to work after childbirth, but to the same level and type of job that they held before becoming mothers, then many of the unfavourable consequences often associated with taking time away from work for children may be avoided. If employers take steps to facilitate the obvious wish of many women to continue working during family formation, and do so in ways which enable women to make continued use of their skills and expertise, then at least some of the problems associated with skills shortages and staff retention also may be avoided.

Returning to work after childbirth

It was not until the mid-1980s that company managers turned their attention to the problem of retaining women in employment during family formation. Companies such as Esso, ICI, and British Gas, who for many years had employed substantial numbers of women in clerical occupations, realised for the first time that their female retention rates in professional and technical positions were poor in comparison with those of men. Both ICI and British Gas discovered that a high proportion of women going on maternity leave did not return to work with their companies; indeed, up to one third of such women in ICI normally failed to return from maternity leave. One manager described what happened at Esso, a company that had recruited a relatively large number of women into higher-level occupations in the 1970s and 1980s:

> We started looking at data in very fine detail and found that the turnover rate for female MPTs (manager, professional and technical grades) was very high. It followed a standard pattern which was for women to join the company, stay around for a few years, start families and leave the organisation. It has only been since the mid-80s, when we started to focus on the whole area of women and men and families, that we have actually started to turn the figure around.

The low retention of women in professional and higher-level jobs experienced by these companies was not unusual and tended to reflect the pattern of women's employment generally. Nevertheless, the realisation that valuable human resources were being lost prompted management, in these and other companies, to consider actively the introduction of policies which might encourage women to return to work after childbirth. In four of the companies we studied – ICI, Esso,

Boots and British Gas – this led to the development of extended or enhanced maternity leave schemes and the introduction of career-break schemes. Ove Arup and Marconi Defence (Stanmore) also had brought in career-break schemes, and STC Telecommunications was exploring a range of options for retaining women including career-break schemes, childcare provision and flexible hours. According to one manager, the aim of such policies would be

> to make the transition more comfortable, if one can, such that people [women] are prepared to come back and try it. If people come back and try it, you have a reasonable chance of their carrying on. It's the not-coming-back which is the problem.

The development of policies to encourage women to return to work was not even, neither across the companies we studied nor within individual companies. In some instances, policies existed on paper, but had not yet been put into effect in practice other than on an ad hoc basis. Indeed, it was common for the employees we interviewed to say that they could not comment on their company's maternity or childcare policies because they did not know any women with children working in the company, or because they did not know any women who had returned to the company after having a family. An employee from one company reported that:

> Certainly up to five years ago, most women who had families did not come back. Now there's one or two who have come back... (018, female).

Moreover, there was very little evidence of any trend towards more flexible employment patterns such as jobsharing and homeworking among the companies we studied. Jobsharing was the least mentioned and least favoured option for combining work and family commitments among the employees we interviewed and was considered operationally not possible by managers. Very few employees expressed an interest in homeworking. Those who did so were more likely to be software engineers than in other fields. Managers tended to be less sceptical of homeworking than they were of jobsharing. A few companies, those engaged predominately in engineering in particular, were exploring homeworking as a possible way of retaining women. Plessey and STC already had a small number of women working for them at home. But managers did not foresee any significant growth in homeworking arrangements, and considered that it would remain a transitional form of employment for a small number of women leading to their eventual resumption of work on-site.

Help with childcare

At the time of our study, none of the companies provided workplace nurseries or financial assistance with the cost of childcare, although the issue was being considered in all of the companies. Managers were concerned not only with retaining existing staff but with finding ways of attracting new women employees as well. They were aware that women are an underutilised source of labour supply and that providing help with childcare would be one way of attracting them into the labour market. A member of the Westland's personnel team concerned with employment trends in the 1990s explained:

> One of the big issues of retention is creches and the ability of women to return to work. It must be made easier for them to do so. Creches may also be an important factor in attracting people too.

It seemed likely from our interviews with managers, however, that corporate provision of on-site nurseries or creches would remain a low priority, at least in the near future. Most managers suggested that their companies would be more favourably disposed towards contributing to the cost of childcare which was organised by local authorities, for example, rather than be willing to provide facilities on-site. The cost of on-site nursery care of course concerned the companies we studied. But as yet, managers tended to be unconvinced that the demand for workplace nurseries was sufficient to justify their cost, particularly if the use of such facilities was restricted to women in professional, technical or higher-level jobs.

But the most common argument against the provision of workplace childcare facilities concerned the practical difficulties individual companies would encounter when implementing their own schemes. A wide variety of questions easily came to managers' minds when asked about establishing workplace nurseries. These concerned fluctuating levels of demand from employees, differential demand in different regions, the number of places required and the need for qualified staff to run an on-site nursery, and so on. From management's perspective, the provision of childcare facilities on-site would be a substantial investment yielding only limited benefits.

Many of the managers we interviewed tended to conclude, therefore, that individual companies acting alone ultimately could provide only a patchy service which would be of limited use to both the companies and women. A collectively-organised system of childcare provision would, however, provide systematic support. One manager explained:

> If there is anything that must be led, it is that. The government does not want anything like that at the moment. It has all got to be business-led or anything-led but not the government. I do not think business would mind paying as long as there was some coordinated way of dealing with it. If not, it will be inefficient and ineffective, which is probably the way it will be.

The majority of the employees we interviewed recognised the validity of managerial concerns as regards the costs and practicalities of offering nursery facilities on-site, but believed that the problems were surmountable. Some argued that childcare facilities would pay for themselves by improving the retention of women in key jobs. Others suggested that problems of demand could be resolved by offering places to employees throughout the company, including both men and women. Indeed, a number of employees resented the way in which the issue was discussed as a problem which faced only women. Two senior women employees made the following comments:

> I wish they wouldn't just see it as a perk for women. I think it should be seen as a perk for both parents. They still see the role of childcare falling on the mother rather than the father (005, female).

> Site-based childcare isn't just a thing that they need to provide for women employees. Men employees have young families too, and sometimes they go through periods with sickness in the family or changes in family life. I feel that it's a service to all members on-site and not just to women employees (031, female).

Their views were shared by some of the men employees we interviewed, who also were attracted to nursery facilities as a way of helping their own wives back to work:

> I think that the question of childcare facilities is something we should be considering far more actively. I think a facility, whereby the company sponsors or provides some form of creche which can be used by employees on a discretionary basis would be a tremendous asset in attracting people, maintaining them and motivating them. If I knew that my daughter was downstairs or at the end of this building whilst my wife was at work somewhere, it

would be very comforting to me. You just want them to be near (094, male).

But not all of the women we interviewed wanted workplace nurseries. Many would prefer their companies to give financial help with childcare, either through a childcare voucher system or through personal allowances to employees. This would allow parents to make their own childcare arrangements and avoid the need to bring children to the workplace. Flexibility in working hours also was suggested as an alternative to workplace nurseries, and one that would be more suitable as children approached school age. Rather than seeking one solution to the problem of childcare, then, employees wished to see a range of options that would allow them flexibility in choosing how to combine their work and family responsibilities.

Maternity leave

All women in employment who meet certain qualifying conditions have a statutory right to eighteen weeks maternity pay after stopping work to have a baby whether or not they intend to return to work. In addition, all women with the necessary qualifications have a statutory right to return to work with their previous employers at any time within twenty-nine weeks after a birth. On the expectation that enhanced maternity benefits might act as an incentive to women to return to work, a small number of companies have developed policies that extent the statutory minima. Of the companies we studied, ICI provided an example of the way in which women's statutory entitlements may be enhanced both comprehensively and flexibly. ICI has extended the twenty-nine week reinstatement period and offers women maternity leave for a period of fifty-two weeks. Women who return to work before the end of this period may choose to work part-time during the first six months of their return. They may, in addition, take up to ten weeks additional unpaid leave within two years of their return.

As in other companies with similar arrangements, enhanced maternity leave provisions in ICI were available only to women in professional or higher-level occupations who had the potential for further development in the company, including women scientists and engineers with key technical skills. It was clearly in the company's interest to retain these women, and to encourage them to return to work relatively quickly after childbirth. In the case of ICI, only those women who returned to work within one year after a birth were entitled to a flexible, part-time return. Boots operated its scheme in a similar way, encouraging professional women to return to work earlier rather than later, with the promise of flexibility in their working hours upon returning. Esso and British Gas offered enhanced maternity pay to women who returned to work within a specified period after a birth. At Esso, this bonus payment amounted to two months full pay and was available to women in professional and higher-level occupations within the company. In British Gas, women were offered half pay for twelve weeks; uniquely among the companies we studied, all women who had one year of full-time service with Boots eleven weeks before the baby was due were entitled to the payment upon returning to work.

Enhanced maternity leave provisions have been developed in these companies as a way of encouraging women to return to work following the birth of a child. They have been devised, moreover, in ways which minimise the length of time women will spend away from work. As one manager suggested, the company view *will always be to maximise the number of hours that a person works and the minimum hours they take off to have a baby*. Our interviews with employees offered little evidence that women objected to an early return to work after having a baby. Many of the women we interviewed reported that their preference would be to return to work after a limited period of maternity leave, provided that their return could be staged in ways that allowed a gradual transition from part-time to full-time working hours. Career-break schemes, although heralded by many as the answer to the problem of retaining professional women in employment was, in practice, favoured by only a very few of the employees in our study.

Career-break schemes
Career-break schemes allow women to leave full-time employment for a number of years to raise children while maintaining links with their employers. Such schemes have enjoyed enormous popularity as a way of retaining women, and the setting up of new schemes is widely reported in newspapers and journals. The financial sector and, in particular, the major UK banks were the first to introduce career-break schemes, with companies in manufacturing and production increasingly implementing similar initiatives. Six of the companies we studied – Ove Arup, Esso, British Gas, ICI, Boots and Marconi Defence – had formal career-break schemes. Boots' scheme was the

earliest to be introduced in 1985, with the other schemes coming into place between 1986 and 1988.

All of the schemes in the companies we studied were open to certain categories of employee only. These were, for the most part, men and women with professional, managerial or technical qualifications or those with key skills in shortage areas. All eligible employees were expected to have performed well in their jobs. The scope of company schemes varied. The minimum duration of a career break was between one and two years, and in some cases a maximum break of five years was specified. This maximum could not be exceeded, and was expected to cover the needs of women (or men) who took more than one break. Procedures within companies for the administration and implementation of career-break schemes also varied, but in all companies it was stipulated that employees on career breaks must resign from the company for the duration of their absence from work.

Managers stressed the importance of maintaining contact with employees on career breaks. All of the companies had a variety of arrangements for employees to maintain contact both with the company and with their former departments. Women on career breaks were sent the house journal and other company communications, for example, and were expected to work for the company – either on-site or at home – for a short period during each year they were away. From the company's point of view, keeping in touch with their employees was considered vital to the success of their career-break schemes and would allow women to resume their careers as quickly and easily as possible.

All of the companies encouraged women on career breaks to remain in touch with developments within their profession. Managers felt that re-entry training would be minimised if effective contact with the company and skills were maintained throughout the period of the career break. Accordingly, managers reported that women were encouraged to attend briefing or training sessions at the company's expense. Women were also encouraged to obtain further qualifications while on a career break. Marconi Defence, Stanmore, for example, paid a retainer fee for up to five years to any women on a career break as a contribution to annual institute subscriptions, travelling expenses for conferences and the like. A manager at Marconi explained that it was in the interest of both employer and employee for skills to be maintained:

> We see it as being very important because it is a high technology environment. If people lose contact, they are also losing contact with the experience that they have gained because things change so rapidly. It's very important from their point of view. We also wish them to be productive when they return to work.

ICI offered a guaranteed job upon an employee's return from a career break. Company policy at ICI dictated that efforts would be made, where possible, to ensure that the post would be at the same level, of the same nature and in the same department as before the career break. This commitment was made to ensure *that an individual's career should not be unduly disrupted as a result of their break.* The remaining companies also stressed that every effort would be made to re-employ an individual who had taken a career break and to do so at an appropriate grade, but offered no specific guarantee.

Career-break schemes have been favoured by managers because they are inexpensive and relatively easy to implement. Esso was one of the first companies in the manufacturing sector to introduce a career-break scheme. A company manager explained the popularity of the scheme within his organisation:

> It's a good example of something we felt that we could put into place successfully and which would not present any difficulties for the powers-that-be in terms of managing the company. It was an acceptable policy for all those sorts of reasons. It was seen to be something which would demonstrate very early on the sort of nature of our commitment. It was a very good policy initiative to take up from that standpoint.

All of the employees we interviewed welcomed the introduction of career-break schemes in the engineering industry generally. They saw the scheme as an asset which would allow women more time off when they became mothers while simultaneously providing a way for women to maintain their place in the labour market. Nonetheless, many of the younger women we interviewed, particularly those without children, considered that they personally would not take a career break after having a baby, but would return to work following a limited period of maternity leave. They had serious reservations about career breaks because they were largely untried and untested. Their worries focused on maintaining contact over the break, their place in the company after returning to work and their future career prospects. A women structural engineer noted:

> They haven't really begun to address the problems of what do they do with these women who come back. What are their promotion prospects? Are they lumped with the people who are five years behind them? Are they marked down because they didn't come back straight away like some other women who did? Nobody has talked about that at all. It's just assumed that you go off and come back. But come back to what? (044, female).

The career-break schemes in place in the companies studied were relatively recent innovations, and had not yet been widely used. No internal evaluation of their effectiveness had taken place. At the time of the interviews, only a handful of women had taken a career break in each of the companies. Very few women had returned to work after such a break. All of the companies had ideas about amending company policy and some were already informally 'breaking the rules'. As a manager at Marconi Defence, Stanmore, suggested:

> We are looking at ways of being more flexible with the scheme. At the moment you have to be employed by the company as an engineer for five years before being eligible for the scheme. So we are looking at that criteria. We look at each case on its merits because I think it is very important that people see that we are trying to help them. We are not putting up barriers. If someone has been here four years, we are not going to say sorry for the sake of it. That is quite important.

Our evidence suggests that there is a need for flexibility in the development of career-break schemes. Furthermore, it appears that flexibility in working hours is more important for women with young children than career breaks. Boots, for example, had found that few of their professional or senior women wished to take an extended break from work. Maternity leave with a flexible return to work was the preferred option among its professional women employees. Indeed, the wish for greater flexibility in working arrangements was an issue raised by men and women alike. Employees from all of the companies spoke about informal systems of flexi-time which operated in some departments but not others. Almost all of the employees felt that their companies, or at least their local managers, were flexible when it came to domestic or other emergencies. But they wanted to see opportunities for flexible hours to be introduced on a regular basis, and in ways that were not dependent upon the good will of individual line managers. They wished in particular to have the opportunity to return to work on a part-time basis.

Part-time employment

The managers we interviewed reported that their companies were finding it more difficult to implement part-time working arrangements for women in senior or professional jobs than had been the case with career-break schemes. Few had experienced part-time employment at any level within their organisations, much less at technical or senior levels. One manager remarked:

> We have never embraced part-time working. The life-blood of this organisation are operations that run for every hour of the year. We have no experience of it. We have no knowledge of it. We do not know how to manage it. We do not know what the consequences are.

Managers emphasised the extent of change that would be necessary to introduce part-time employment in senior or technical positions within their organisations. They stressed that it would raise fundamental questions about the organisational structure of the company. It would require changes in the design and specification of jobs, and would raise questions about supervision and the ways in which people work together. Managers felt that continuity between the work of team members, for example, would be difficult to maintain. Employment benefits, such as holiday and pension entitlements, would have to be reassessed. The career structure of the organisation would have to be redrawn in order to encompass both full- and part-time careers.

Consequently, most managers thought it unlikely that there would be an increase in part-time working opportunities for women in senior or technical jobs, at least in the near future. They could not imagine changes in working practices which would make part-time employment possible and argued that middle and higher-level managerial and technical jobs were not compatible with less than full-time hours. The difficulties rather than the possibilities were emphasised:

> If you think about senior management positions within the company, I think it would be extremely difficult. They could not really be done on a part-time basis. It would not be a feasible role to carry out two days a week while everyone else was working five.

> I think we have got a fairly limited ability to offer it because of the nature of the work. If we were a computer company or something like that the opportunity to so that sort of thing would be much greater but here the physical requirements are that people have to

be in the labs working because they work as a team which would make it difficult.

There has to be a sensible agreement between the company and the person. One could not guarantee, for example, that someone would be able to work half their normal hours in any job. That just is not physically possible.

Similarly, one of the employees we interviewed who also had managerial responsibilities argued that:

It is extremely difficult to do the sort of research that we do part-time. It has almost always been the case that when women have become pregnant they have either decided to come back full-time or not at all because we don't entertain part-time working. That comes down to the sheer cost of maintaining laboratory overheads. They have to be fully utilised (098, male).

But despite speaking at length about the different types of problem associated with part-time working for women engineers and scientists, managers in all of the companies we studied reported that they were exploring ways in which opportunities for part-time employment in technical jobs could be created. One of the most compelling reasons for exploring the possibilities for part-time employment was that women returning from maternity leave increasingly wanted to resume work on a part-time rather than a full-time basis. Companies had turned down such requests in the past but were now having to rethink their ideas. One manager remarked:

Women would prefer to come back part-time rather than take a career break. The career-break option is not all that attractive. If you are really trying to pursue a career and you are serious about wanting to keep in the career stakes, you want to keep involved. A career break is quite a break. You are walking away from that.

Accordingly, we found examples of both formal policies and ad hoc arrangements for part-time employment. British Gas, Boots and ICI, as we have noted, had formal arrangements in place which allowed women returning from maternity leave to resume work on a part-time basis. Early opposition to part-time employment within British Gas had given way to support as its skills retention package was implemented in the different regions. As one manager explained:

It went down like a lead balloon [at first]... But it is happening because you get someone who is very good and you look at what is out on the open market and you think 'I will have half of what I know rather than double of what I do not. There is the possibility that as time goes on that you can perhaps negotiate more hours and bring the person back full-time eventually. So that was difficult at first but in practice has gained support.

Ad hoc arrangements were found in Esso and Ove Arup and were common among the predominately engineering companies. A number of individual arrangements for part-time had been negotiated, for example, at Plessey sites. A manager at their Roke Manor site explained:

> We have been a bit more flexible in our approach. We have got two people to come back to work recently who have both come back initially part-time. They have got the key skills and I think employers generally are being much more flexible about ways of accepting women back.

STC also employed a small number of women part-time and had recently included them in the company's pension scheme, contrary to past practice.

The majority of women employees we interviewed felt that there should be opportunities for women like themselves to work part-time. While acknowledging the difficulties of combining part-time employment with team work or project work where demands varied, most felt that their work tasks and job specifications could be adjusted to accommodate part-time employment. But they also felt that, for themselves, part-time working was not a genuine option. Where part-time opportunities were available within their companies, they were confined to lower-level occupations. Many women felt that their employers were against part-time employment. They believed it was the view of management that part-time employment would not be appropriate for their type of company or their type of job. Accordingly, they argued that the potential contribution women working part-time could make went unrecognised.

Moreover, our interviews with employees suggested that because most arrangements for women to return part-time had been made only very recently or on a piecemeal or ad hoc basis, some women had to push their site managers very hard to reach an agreement on reduced hours, and their eventual return to work had not gone smoothly. One woman who worked part-time for only a short period described the problem in her company:

> I did want to work part-time. I mean I would like to work part-time now. In theory I can do that but in practice it's still not working like that because they haven't changed the manpower figures.

> They've allowed managers to have part-timers on their books permanently but they still count for one manpower person. Two part-time people would be two manpower numbers. So any manager who has got only a limited resource in terms of manpower to get a certain amount of work done is going to want to take a full-timer in preference for a part-timer because they're wasting half a person. So although they've changed the rules, they haven't changed the rules in practice.

One woman had been asked to return on a self-employed basis in order to work part-time, and she was uncertain of her status within the company:

> When I first left to have my first child I was a senior mathematician but when you go part-time all sorts of problems arise and they don't know what to call you. There is definitely a problem in continuing with promotion when you go part-time... I don't know what my official title is.

Another women engineer who had decided to return part-time after the birth of her first child found that no preparations had been made for her return:

> My manager at the time had actually sketched out a rough job description of what I would be doing and there we left it. I came back to find they'd reorganised the section and had done very little about sorting out my contract, which was a bit disappointing.

Arrangements for her reduced hours of work and rate of pay largely had been forgotten in the reorganisation, leading her to conclude that *the intention was there but the arrangements were less than ideal.*

The impact of career-break schemes and part-time working on women's promotion prospects

The large majority of men and women employees thought that taking a career break or working part-time after having a baby would harm a woman's promotion prospects. Both were seen to demonstrate a reduced commitment to the company. A career break, for example, would mean that the employee would have spent several years devoted to other commitments before returning to the company, and it was considered to be unfair for that employee to be treated as the equal of someone who had remained continuously employed:

> Why should someone who has worked for ten years, had five years off, come back and enjoy the same benefits as somebody who's

been there fifteen years? If you look at it from the company's point of view there can be no doubt which one you can expect to be pushed forward (023, female).

Career breaks and part-time working affected managerial perceptions of an employee's priorities; they affected the employee's length of service and their accumulated work experience. Therefore, many employees expected that taking a career break would slow down rates of promotion:

> If they take advantage of some of the policies on career breaks or part-time, then there is no doubt that that must impact on their careers. As it would do with a man... you need to have a workload behind you in order to be able to move on to the next level (092, male).

Career breaks in particular were seen as problematic. One woman we interviewed reported that she was still confident about her promotion prospects *unless I decide to go on a three year career break* (042, female). Another said that it had taken her *a good year* to re-establish her position and responsibilities on returning from a career break (029, female). One employee from Esso had returned, however, to a more senior position after her career break.

Women choosing to work part-time because of family commitments also had experienced problems in attempting to advance their careers. They expressed, at the same time, dissatisfaction at the damage done to their careers and gratitude that the companies had allowed them to return to work part-time. The experiences recounted below are typical:

> Since I came back I've been demoted or gone down a grade... when I left I made it quite clear I didn't want to come back to full-time employment and the managers at the time felt that they couldn't offer me part-time employment at the same grade. A part-time employee would have to be doing a lower grade work which wasn't quite logical to me... to balance that I was very grateful that [the company] actually employed me part-time because they were under no obligation to (030, female).

> I know one person who has come back part-time but I also know that person has literally been told you can work part-time but don't expect to go up the ladder. You either come back full-time and go up the ladder or you work part-time and stay where you are (042, female).

These are, of course, the reports of only a few individuals, but they reflect problems that are likely to exist on a wider scale. Career-break schemes were introduced largely because they are easy to administer, relatively cost-free and provide good public relations in terms of equal opportunities. In practice, however, few women in professional and technical jobs want to take breaks from work lasting one, two or more years. Our evidence suggests that such women would prefer a period of maternity leave followed by a phased return to work, gradually increasing their hours of work from part-time to full-time. Arrangements for part-time employment in technical or senior jobs have, however, proved very difficult to implement. Indeed, few managers had addressed the issue of part-time employment and career progression. While women who took maternity leave or a career break might fall behind colleagues who had not been similarly absent from work, they could still resume the career ladder on returning to work full-time. The potential for combining part-time and career progression was not so obvious.

Notes

1. Stopping work for childbirth has been shown to have unfavourable consequences for women's subsequent labour force participation. This is particularly the case for women who return to work on a part-time basis. Joshi and Newell (1987), for example, have shown that women incur substantial losses in lifetime earnings as a result of breaks for childbirth. These losses are both direct in that they derive from women's absence from the labour market; and indirect as a result of reduced earnings through entering lower-paid part-time employment after childbirth. Women who return part-time are likely also to experience downward occupational mobility. Martin and Roberts (1984) found that 45 per cent of women who returned to work part-time returned to a job at a lower occupational level, compared with 19 per cent of women who returned full-time. See also the discussion in Chapter 5 of the present report on the consequences of children on women's opportunities for promotion.

2. In order to qualify for Statutory Maternity Pay, a woman must be continuously employed with the same employer for six months ending in the fifteenth week before her baby is due. In addition, her earnings must be sufficient to attract National Insurance contributions. The requirements for the right to return extend the continuous employment period to two years for women working more than sixteen hours each week, and five years for women whose weekly working hours are between eight and sixteen hours.

3. In most companies career-break schemes are open to both men and women although the very large majority of participants in such schemes are women.

4. Boots, a manufacturing and retailing company, offered senior-level part-time employment, but largely confined such arrangements to the retailing arm of the company. A manager within Boots explained:

> We have no problem finding a part-time position for pharmacists because the retail trade is so geared up to part-time working that we can organise the hours to suit the individual and the management and the store accordingly. So there is no real problem for a women who is on maternity leave and has asked to come back part-time. We can sort that out in 95 per cent of cases. The manufacturing area is a little more difficult.

7 Conclusions and recommendations

In this chapter we summarise the main findings of the study. Our research has been based upon the policies and practices of ten major British companies. However, we believe that the results of the study extend beyond these ten companies and have value for other companies, individuals and policy-makers concerned with improving the supply of engineers and scientists to industry. Accordingly, we draw out the study's implications by identifying examples of good practice among the companies we studied and by specifying the steps that companies in general would benefit from taking, rather than focus upon potentially beneficial changes in policy and practice in relation only to the companies we studied.

Schools liaison
Schools liaison projects attempt to encourage young men and women to study technical subjects like physics, chemistry and mathematics at GCSE level, at A level and beyond. Managers reported that schools liaison policies were becoming increasingly important for their organisations. From their viewpoint, schools liaison was a long-term recruitment strategy. With the appropriate technical qualifications and skills, such young people would constitute one part of the pool of labour from which the companies could draw in the future. Managers recognised the difficulties of measuring the effectiveness of their schools liaison policies. Nonetheless, there was widespread agreement that such projects did have beneficial effects, albeit ones that were small and as yet, unmeasured.

Conclusions and recommendations

Perhaps the main aim of schools liaison projects was to dispel the image of technical careers as involving greasy, rather dull work done in dirty overalls. To overcome this image, employees would try, through a variety of projects, to show pupils that technical subjects were interesting and fun, as well as being intellectually challenging. The importance of technical subjects as springboards to a wide range of careers was also stressed, together with the practical applications of technology, especially in non-military areas: gas fires, hospital equipment and cookers rather than off-shore oil rigs and tanks.

These activities were seen to be the most effective when they were carried out in the same schools in the local area over a number of years. Several of the companies we studied had established a series of linked schools. Some had as many as fifty links with specific schools; others had only one or two. Concentrating schools liaison activities in schools nearby allowed companies to build up good working relationships with teachers and to develop familiarity with the problems of particular schools. This was seen to yield cumulative benefits to both schools and companies over time.

Schools liaison did not stop, however, with employees going into schools. In addition, pupils were brought into companies through work placements and teachers through work shadowing. Work placements were found to be most effective when they provided genuine training for young people. Rather than simply giving pupils something to do to fill their time, companies were now attempting to give them real work, genuine training that might well convince them to follow a technical career path. Work shadowing was seen as important as a way of overcoming teacher ignorance and prejudice. Few teachers know much about industry, although many have fixed ideas that it is not where they want their brightest pupils to go.

Schools liaison projects appear to act as a positive, but limited, influence on the educational choices of some pupils. The limits to the effectiveness of links with schools were found to be both quantitative and substantive. Companies had contacts with only a small number of pupils and teachers; furthermore, employees involved in schools liaison and managers alike reported that the pupils and teachers with whom they had contact were often self-selected and already receptive to the idea of technical careers in industry. Talks and activities were greeted with enthusiasm by pupils, but seemed ultimately to have little impact on the subject and career choices of the majority. Teachers were equally enthusiastic, but company representatives seldom met the other, less-well-disposed teachers. At career talks, only those parents who were already willing to see their daughters or sons enter technical careers came forward. The others stayed away.

One implication of this is that schools-industry links would benefit by an extension to primary schools. Stereotyped views of industrial careers, particularly in relation to women, were found to be persistent. By the time boys and girls have begun their secondary education, many have formed negative images of technical subjects and careers in the technical field. It appears likely that, to be fully effective, schools liaison projects will need to be directed at much younger children.

Our study suggested further that a need exists for young people to be given much more detailed advice about the range of degree courses on offer at universities and about the vocational implications of their educational choices. This appeared to be the case particularly in relation to careers advice concerning engineering. The evidence from our interviews with employees suggested that more frequent and structured contacts between industry and schools need to be established before positive attitudes towards technology would be reflected in technological career choices.

Our study suggested, however, that in the absence of quantifiable indicators of success, schools liaison projects may be vulnerable to financial cutbacks. It appears to be in the interest of companies to develop ways to measure the benefits of schools liaison and to monitor as closely as possible the results of their activities.

Good practice
Our study allowed us to identify the following examples of good practice in relation to schools liaison projects:
1. concentration of corporate activities on specific, linked schools in the local area, which allowed companies to build up expertise and familiarity with the problems of the schools over a number of years and yielded cumulative benefits to both schools and companies;
2. emphasis on the practical application of technology, especially in non-military areas, which allowed pupils and teachers to gain some understanding of the relevance of industry to their everyday lives;
3. use of work placements for pupils to provide genuine training;
4. emphasis on the intellectual content of technical work as a way of dispelling its image as dirty, manual work;

5. emphasis on the importance of technical subjects as a springboard to a wide range of careers in industry;
6. provision of special training for employees engaged in schools liaison projects, in recognition of the benefits which accrue to employees as well as to pupils from such activities;
7. collaboration with educationalists and others to share expertise and practical knowledge;
8. provision of work shadowing and company visits for teachers.

Next steps
Our findings as regards schools liaison projects suggest that companies would benefit from the following:
1. introducing or extending work placements for pupils, and ensuring that the work done on such placements is challenging and meaningful to pupils;
2. introducing or extending work-shadowing arrangements with teachers, and working with teachers to change the ways in which science is taught in schools, with greater emphasis being directed at non-military and non-industrial applications of technology;
3. directing schools liaison projects towards younger students, in particular, towards pupils in primary schools;
4. finding ways of measuring the success of schools liaison projects.

Graduate recruitment
The companies we studied favoured the recruitment of women. They wished to project an image of themselves as equal opportunity employers, and gave women scientists and engineers a high profile wherever possible. They had adapted their graduate recruitment to indicate their willingness to recruit women into technical positions and to ensure that women were equally considered for such jobs.

The strategies they had adopted included giving women a high profile at career talks and fairs and in corporate recruitment literature. Most companies included details of the careers of women in their *employee profiles*. Women were also used to interview potential recruits, particularly in the first, milkround, interviews with students. The majority of companies were also trying to include women in on-site interview panels, but often came up against the problem of having too few women in senior positions.

Monitoring was another strategy. Most companies monitored their overall graduate recruitment activities. A few had begun to monitor this process with a view to tracking their employment of women graduates. These companies were anxious to see that their recruitment of women matched national figures on the proportion of women graduating in science or engineering.

However, despite the efforts which companies put into the development and implementation of equal opportunities policies, success in this area could be undermined by the attitudes of individuals in positions with the power to influence recruitment decisions. Reports of bias and prejudice among personnel with recruitment responsibilities were made by employees in all of the companies we studied. Men and women employees from many of the companies maintained that all else being equal, they – or others in their companies – would prefer to employ men rather than women. This position was justified by reference to women's traditional responsibility for the care of their families and to the potential conflict between work and family commitments. In addition, women were seen as unsuited to engineering work, or as inappropriate simply because they were women.

The interviews we carried out with management suggested that in all of the companies we studied, steps had been taken towards formalising recruitment procedures in order to eliminate this type of discrimination against women and to ensure equal opportunities between men and women. Some companies were further along this path than others, but all were engaged in the process. As part of this, a number of companies provided extensive training in interview techniques, including equal opportunities training. This did not always mean that all interviewers received training or that training was compulsory for people carrying out interviews. Moreover, in most of the companies, general interview training and guidance with specific reference to equal opportunities was less systematically provided to on-site interviewers that to those who carried out milkround interviews. It was often left to the discretion of line managers or senior interviewers to apply for a course on interviewing techniques. Companies could not, therefore, guarantee that all senior employees engaged in interviewing had received interview guidance training and could be even less confident about specific training concerning equal opportunities policies.

Two points stand out very clearly in this regard. First, the success of equal opportunities policies relies on the commitment of all employees. Wide-ranging consultation with employees is vital for such policy to be put into practice by people at all levels within a company. Secondly, the implementation of policy requires the active support of employees with line managerial responsibility, in order that new procedures are followed. Implementation of these two conditions will go some distance in ensuring that the aims of an equal opportunities policy will be fulfilled.

Good practice
The companies we studied had adapted the graduate recruitment process to signal their wish to recruit more women. Among the steps they had taken the following appeared particularly effective:
1. highlighting in recruitment literature the intellectual and challenging aspects of technical careers and the opportunities for varied career paths;
2. highlighting the careers of successful women employees in recruitment literature through the inclusion of employee profiles;
3. using younger employees, especially younger women, in careers talks, fairs and conventions;
4. including women in both milkround and on-site interview panels;
5. undertaking detailed monitoring of graduate recruitment as part of a central administrative process to track trends in the employment of women and men;
6. using the results of monitoring to measure the performance of interviewers.

Next steps
It appears from our study that companies would be likely to benefit from steps to formalise their recruitment procedures and, in particular, by implementing equal opportunities training for senior and line managers. Companies also appear likely to benefit from the following:
1. centralised recruitment and dual interviewing with a personnel division interviewer trained in equal opportunities;
2. equal opportunity training for all interviewers, including employees in senior posts and employees involved in on-site interviews and assessment panels;

3. the development of recruitment profiles (person specifications) for use by interviewers so that they will be less likely to be influenced by subjective attitudes towards applicants;
4. ensuring that at least one woman was included in every interview panel, bringing in expertise from personnel or other departments if necessary.

Training

British companies have been criticised for providing insufficient training for their employees. At least four of the ten companies we studied appeared to be responding to such criticism, with employees reporting that training had improved in recent years. In all four companies, the situation had been improved by the introduction of more-formalised training programmes.

Many of the employees we interviewed said that they had decided to join relatively large companies because they offered the promise of a thorough training. Younger employees in particular saw training as an essential part of career development, along with the opportunity to acquire a breadth of experience. It is important that training opportunities are not only available but also seen to be available. Individual mangers may be in a good position to offer advice on specific courses that would be useful for employees in their sections. Our evidence suggests that formalised and centralised systems of training allocation may reduce variations in availability which tend to flow from a reliance on the goodwill of individuals. In cases where more-formalised systems had been introduced, they were generally viewed as bringing improvements in the availability of training. Most employees favoured a more structured system with clearly-specified training entitlements and opportunities.

The large majority of managers and employees believed that men and women had equal access to training opportunities, but there were some areas of provision, such as training opportunities for part-time workers, which still give cause for concern.

Good practice

In the area of training, the following practices stood out as particularly beneficial to companies:
1. individually-tailored training programmes, agreed by the individual concerned, which deliver job-related rather than general training;

2. well-organised induction courses, delivered shortly after the new recruits has joined the company.

Next steps
Our findings as regards training opportunities suggest that companies might benefit from the following:
1. a reappraisal of *individually-determined* programmes of training, with a view to establishing more-structured and more-formalised systems of training for new graduates in particular;
2. increased opportunities for training leading to managerial opportunities.

Career progression
In most of the companies we studied promotion opportunities were relatively automatic in the early years of a science or engineering career. This provided a good sense of progression for young employees. Our interviews with employees indicated that many of them were prepared to leave their companies if they felt that promotion was too slow or that opportunities for promotion were too limited.

Opportunities for promotion became limited as employees moved up the hierarchy of seniority. They were restricted also in smaller companies or departments with few vacancies in senior posts and where there was little mobility between functions and departments.

Opportunities for promotion also were limited for employees who wished to pursue purely technical careers. Technical career ladders existed, but in most companies they tended in practice to be shorter and narrower than managerial ladders. Technical careers were perceived to bring less recognition and inferior rewards.

Procedures for assessing promotion had changed in recent years in many of the companies we studied, and in most companies, continued to evolve. Women stood to benefit from these changes and indeed, some had been introduced in order to increase the likelihood of retaining women employees. The strategies we identified as particularly important included the introduction of objective or formalised assessment. Most companies used some system of formal assessment or grading to structure career progression and the majority of employees favoured formal systems based on explicit criteria. The introduction of such schemes encouraged employees to set annual targets of achievement and provided avenues for the discussion of non-achievement of these targets. Many companies had introduced individualised career paths into their management planning, whereby managers and employees worked together to decide where an employee wanted to go in his or her career and how best to get there. Dissatisfaction existed when such systems existed in name only, and where managers did not adhere to the rules set down.

The monitoring process had also been extended in some instances into the area of career progression. Few of the companies had any women in senior managerial positions; none had a women board executive. All stressed that the scarcity of women in their top jobs was a consequence of the only very recent entry of women into careers in technical fields. It was suggested that many young women were now progressing through the lower levels of management and that, in time, they would occupy senior positions. Accordingly, it was seen as important to monitor the performance of women over the next few years to assess whether they were in fact gaining the promotions that might be expected. It was believed that monitoring the promotion procedures and mobility of men and women would lead to greater equality of opportunity; in addition, monitoring could provide the statistical evidence of unequal opportunities useful to managers in their attempts to translate policy into practice.

A primary objective behind the introduction of objective assessment procedures was to eliminate discrimination against women. Our study revealed the existence of barriers to women in the technical field which were similar to those found in other parts of the labour market: outmoded attitudes about the role of women; direct and indirect discrimination; the absence of proper childcare facilities; and inflexible structures for work and careers. Whatever problems might exist in an organisation in relation to the recruitment of women to technical occupations, they are compounded when the issue is how to get women into senior jobs. Our interviews with employees suggested that considerable scope remained for individual preferences to be the deciding factor in promotion, and with individual preference comes individual prejudice.

Comments were made about difficulties with customers or clients being confronted by a senior woman manager and about employee resistance to having a woman boss. But the most common explanation put forward for the scarcity of women in senior jobs was that many potentially eligible women left work to care for children and failed to return. The level of commitment expected by some companies put considerable pressure on employees' personal and family lives, and this was widely perceived to make promotion difficult for women.

All women, not just those with children, were said to be put at a disadvantage by corporate cultures which required that candidates for promotion should give first priority to their careers. Unmarried women and women without children were equally affected by the assumption that the family would be a woman's first priority: if not now, then someday in the future. A few of the individuals we interviewed saw all women as inherently unreliable, because of their actual or potential family responsibilities.

The lesson to be drawn from our research is not only that some managers and employees have attitudes which act against the best long-term interests of their companies, but also that the introduction of equal opportunities policies is, by itself, not sufficient. Discrimination against women assumes many forms and may persist despite the best efforts of senior management. Innovative policies designed to help women accommodate domestic demands with commitment to careers, such as career breaks and part-time working arrangements, were seen by some to be detrimental to women's chances for promotion.

Good practice

Our study allowed us to identify the following examples of good practice in relation to career progression:
1. the introduction of formal assessment schemes which encourage employees to set annual targets of achievement and provide avenues for discussion of non-achievement of such targets;
2. involving employees in their own career development plans through assessment meetings which identify potential career paths and take into account the employee's long-term career goals;
3. monitoring the progress of women employees in relation to promotion;
4. providing special training for women with managerial potential.

Next steps

In consideration of our findings, it is apparent that companies would benefit from the following:
1. continuing to improve the career prospects of employees who wish to remain in technical work;

2. where necessary, introducing open and accessible performance-assessment procedures under which performance is assessed against neutral and measurable criteria, rather than a normative profile of skills and personal characteristics implicit in which are judgements about what is acceptable;
3. ensuring that policies designed to guarantee objective assessment are followed in practice;
4. considering the implications for career progression of part-time working arrangements and devising ways in which part-time workers can continue to advance within the company;
5. ensuring that at least one woman is included in every selection panel where promotion is considered.

The retention of women engineers and scientists

In response to problems of labour supply, and acting in accordance with their equal opportunities policies, many of the companies we studied had developed enhanced maternity leave provisions and had brought in career-break schemes for women who wished to leave work for a period in order to have children. Other companies – but only a very few – were considering the feasibility of part-time working for women in senior and technical jobs.

It was clear from our interviews that women and employers wanted the same outcome, which was to find effective ways of retaining women in work after having children. However, the problems they must solve in gaining this outcome were not similarly in accord. For companies, the problem was to balance the need to introduce policies that will encourage valued, well-qualified women to return to work after childbirth, with the equitable demand that all such policies should be economically and administratively feasible. For women, the problem was to balance the wish to maintain satisfying careers and opportunities for advancement in professional and higher-level occupations with the demands, of both time and energy, and the pleasures entailed by having young children.

Our study suggested that, from the perspective of management, enhanced maternity leave and career-break schemes appeared to be the best solutions. Enhanced maternity leave encourages a relatively quick return to work following childbirth, normally on a full-time basis. Career-break schemes are cost-effective, easy to administer and may, over time, achieve the desired outcome.

Many of the women we interviewed tended, however, to prefer a period of maternity leave followed by a phased return to work, gradually increasing their hours of work from part-time to full-time. They wanted more flexibility in both childcare arrangements and working hours, which would allow some choice about the best way to combine work and family responsibilities.

But arrangements for part-time employment in technical or senior jobs have proved very difficult to implement. It was widely believed by management that such jobs do not suit part-time or jobsharing arrangements. These jobs were seen as unsuitable for less than a full-time commitment, with fears expressed about disruption to team-working and organisational efficiency and a loss of managerial control.

In addition, the large majority of men and women employees thought that working part-time after having a baby – or taking a career break – would harm a woman's promotion prospects. Both were seen to demonstrate a reduced commitment to the company. A career break, for example, would mean that the employee would have spent several years devoted to other commitments before returning to the company, and it was considered to be unfair for that employee to be treated as the equal of someone who had remained continuously employed.

Few managers had addressed the issue of part-time employment and career progression. While women who took maternity leave or a career break might well fall behind colleagues who had not been similarly absent from work, they could still resume the career ladder on returning to work full-time. The potential for combining part-time working hours and career progression was not so obvious.

Good practice was more variable among the companies we studied in relation to measures to retain women employees than for other areas investigated. Accordingly, we are less able to draw a clear divide between good practice and next steps: *good practices* in some companies were likely to be in other companies potential *next steps*. In short, the measures introduced in some of the companies we studied need to be introduced more widely, while all companies would benefit from the introduction of greater flexibility in working arrangements.

Next steps
It seemed clear from our study that companies would benefit from continued movement towards flexibility in working arrangements for

women with children. In particular, companies would benefit by offering their employees a range of options, including:
1. enhanced maternity leave, with provision for a gradual return to full-time working hours;
2. part-time working hours or jobsharing arrangements in senior and technical occupations;
3. where necessary, the introduction of career-break schemes as a complement to enhanced maternity leave and other flexible arrangements;
4. provision of on-site childcare facilities where economically feasible and/or the introduction of childcare vouchers. Companies might also benefit from encouraging employers' interest groups to act collectively to improve the provision of childcare in Britain generally;
5. monitoring rates of return from maternity leave, and surveying non-returners to discover their reasons for not returning.

In many ways, our study has highlighted the limits on the influence of employers in attempting to bring about social changes in the attitudes of pupils, parents, teachers and, in some instances, their own employees. Many, perhaps all, of the companies we studied could be seen as *model employers* in relation to the employment of women in higher-level, professional occupations traditionally perceived as appropriate for men. The steps they have taken, and are continuing to take, to encourage women to enter engineering and science and to remove barriers to their continued participation and success in technical careers have not been slight nor cost-free. It is likely that the majority of the companies we studied would argue that they have had many successes. It seems equally likely, however, that most would agree that much remains to be done if women are to be enabled to compete equally with men in technical occupations and if employers are to ensure an adequate, and well-qualified, labour supply.

Note
1. For a report which surveys both the barriers which confront women in gaining access to senior and higher-level positions across a wide variety occupations and the strategies undertaken by employers and others for overcoming the barriers, see *Women at the Top*, Report of the Hansard Society Commission, 1990.

Appendix

Characteristics of the employees who participated in the study

The purpose of this appendix is to provide some details about the employees we interviewed. In all 130 interviews with employee were carried out although detailed personal and educational information is available only for 128 employees. The breakdown of number of men and women interviewed by company is as follows.

Company name	Women	Men
British Aerospace	7	4
British Gas	5	8
Boots	8	7
Esso	7	7
ICI	7	7
Marconi	7	7
Ove Arup	8	–
Plessey	7	7
STC	7	7
Westland	7	6
Total	70	60

Personal characteristics
The majority of women employees were under age thirty-five, married or living as married and had no children. Almost half of the women employees were single. The men we interviewed tended to be older than the women. Proportionately more men were married; they were more likely to be fathers and to have a larger number of children.

Table 1: Age

Column percentages

Age	Women	Men	Total
25 and under	35	22	29
26-35	50	40	45
36-45	13	33	23
46 and over	1	5	3
Base	68	60	128

Table 2: Marital status

Column percentages

	Women	Men	Total
Single	45	18	30
Married/Cohabiting	54	69	66
Separated/Divorced	6	3	5
Total	68	60	128

Table 3: Number of children

Column percentages

Number of children	Women	Men	Total
None	84	52	69
One child	7	12	9
Two children	7	28	4
Three or more	1	8	1
Base:	68	60	128

Appendix

Educational background

Table 4: Degree subject

Column percentages

Degree	Women	Men	Total
Engineering degrees			
Electrical/electronic	23	22	23
Chemical	8	16	11
General	11	8	10
Mechanical	6	6	6
Aeronautical	2	2	2
Civil	2	–	1
Science degrees			
Physics	2	18	9
Chemistry	9	8	9
General	5	2	3
Biology	3	4	3
Botany	3	–	2
Pharmacy	–	2	1
Maths	9	4	7
Maths and physics	2	4	3
Maths and chemistry	2	–	1
Other degrees	10	4	10
Base:	65	50	115

Note: Ten men and two women did not complete degrees; they entered their companies as apprentices and later took OND or HND qualifications. One woman was about to begin a degree course.

Bibliography

Berthoud, Richard and David J Smith, *The education, training and careers of professional engineers*, HMSO, 1980.

Blackstone, Tessa and H Weinreich-Haste, 'Why are there so few women scientists and engineers?', *New Society*, 21 February 1980.

Bruce, M, and G Kirkup, 'Post-experience courses in technology for women', *Adult Education*, Vol 58, No 1, 1985.

Carter, R and G Kirkup, *Women in Engineering: A good place to be?*, Macmillan, 1989.

Collinson, David, *Unfair Selection Practices*, Equal Opportunities Commission, 1988.

Connor, Helen and A Gordon, *Women in Industry: The role of employer sponsorship*, IMS Report No 110, 1985.

Connor, Helen and Richard Pearson, 'IT manpower into the 1990s', *Employment Gazette*, November 1986.

Curran, Margaret, 'Gender and Recruitment: People and Places in the Labour Market', *Work, Employment and Society*, Vol 2, No 3, 1988.

Daniel, WW, *Maternity Rights: The experience of women*, PSI Report No 588, June 1980.

Daniel, WW, *Workplace Industrial Relations and Technical Change*, Frances Pinter in association with the Policy Studies Institute, 1987.

Elias, Peter and Malcolm Rigg, *The Demand for Graduates*, Policy Studies Institute and Institute for Employment Research, 1990.

Else, L, 'Women in Computing', *Computing*, 9 May 1985.

Engineering Futures: New audiences and arrangements for engineering higher education, The Engineering Council, 1990.

Engineering our Future, Report of the Committee of Inquiry into the Engineering Profession, HMSO, 1980.

Freeman C and L Soete (eds), *Technical Change and Full Employment*, Basil Blackwell, 1987.
Girls and Mathematics, The Royal Society, 1986.
Gordon, A, R Hutt and R Pearson, Employer Sponsorship of Undergraduate Engineers, Gower, 1985.
Institute for Employment Research, *Review of the Economy and Employment: occupational studies*, University of Warwick, 1988.
Jenkins C and B Sherman, *The Collapse of Work*, Eyre Methuen, 1979.
Joshi, Heather and M-L Newell, *Family responsibilities and pay differentials: evidence from men and women born in 1946*, CEPR Discussion Paper No 157, 1987.
Martin, Jean and Ceridwen Roberts, *Women and Employment:* A Lifetime Perspective, HMSO, 1984.
Martin, Jean, 'Returning to work after childbearing: evidence from the Women and Employment Survey', *Population Trends*, No 43, Spring 1986.
McLoughlin I and J Clark, *Technological Change at Work*, Open University Press, 1988.
McRae, Susan, *Maternity Rights in Britain: The experience of women and employers*, Policy Studies Institute, 1991.
Northcott J, *Microelectronics in Industry:* Promise and Performance, Policy Studies Institute, 1986.
Pilcher, J, S Delamont, G Powell and T Rees, 'Women's Training Roadshows and the Manipulation of Schoolgirls' Career Choices', *British Journal of Education and Work*, 2, 2, 1989.
Rajan, Amin with Julie Fryatt, *Create or Abdicate: The City's Human Resource Choice for the 90s*, Witherby and Co. Ltd., 1990.
Thomas, Susan, 'How to get WISE', *Times Educational Supplement*, 20 September 1985.
Thomas, K, *Gender and Subject in Higher Education*, Open University Press, 1990.
Whyte, J, *Girls into Science and Technology*, Routledge, 1986.
Women at the Top, The Report of the Hansard Society Commission, 1990.
Women in Engineering, EITB Occasional Paper No 11, 1984.